Ousmane GUEYE

Urban dynamics and land issues in secondary towns

Ousmane GUEYE

Urban dynamics and land issues in secondary towns

The town of Bambey

ScienciaScripts

Imprint

Cover image: www.ingimage.com

This book is a translation from the original published under ISBN 978-620-6-70610-6.

Publisher:
Sciencia Scripts
is a trademark of
Dodo Books Indian Ocean Ltd. and OmniScriptum S.R.L publishing group

120 High Road, East Finchley, London, N2 9ED, United Kingdom
Str. Armeneasca 28/1, office 1, Chisinau MD-2012, Republic of Moldova, Europe
Printed at: see last page
ISBN: 978-620-7-79031-9

Contents

GENERAL INTRODUCTION

According to PAULET J. P. (2006), 87% of the world's population lived outside cities, and the global urbanisation rate was less than 15% in the 1950s, rising to 33% in 1960. [e] On the other hand, according to CIATTONI A. and VEYRET Y., (2013): "*At the start of the 21st century, 50‰ of the 7 billion people on Earth are urban dwellers*". According to projections by the United Nations (UN): "*By 2050, 70% of the world's population will live in cities*"[1] . But the rate of urbanisation remains unequal between developed and developing countries.

This phenomenon of urbanisation, which has long been the prerogative of Western countries, is becoming increasingly important in developing countries and raises many questions today. Indeed, the majority of the world's cities, particularly the major cities or megacities, are located in developing countries (CIATTONI A. and VEYRET Y., 2013). But cities have changed a great deal in recent decades. Both their proliferating peripheries and their crisis-stricken centres (CHALINE C., et al, 1983). The world's cities are growing in size and number, and are becoming major consumers of space. But "*this urban growth is selective and affects the various continents differently*" (PAULET J.P., 2006). As a result, the world's urbanisation rate varies from one continent to another, from one country to another, from one region to another and even from one city to another. It is estimated to be between 73% and 81% in developed countries by 2030, and between 30% and 57% in developing countries (CIATTONI A. and VEYRET Y., 2013).

This urban growth, which is on the one hand the result of huge population movements towards urban areas, will continue, especially in Third World countries, throughout this millennium, as rural population reserves are still very large in Asia and Africa (CIATTONI A. and VEYRET Y., 2013).

In Africa, the current rise in the urbanisation rate can be explained by the agricultural crises in the countryside, with increasingly low agricultural yields causing rural dwellers to abandon their farmland and head for the city. Faced with such situations, the city has become a centre of hope and rejection, attracting people from rural areas for well-known reasons (the search for a better future). The countryside is becoming repulsive because it is in crisis (PAULET J.P., 2006) and is unable to offer rural populations the development opportunities they need to flourish. This tends to encourage the mass movement of people from the countryside to the cities: a phenomenon better known as "rural exodus". In developing countries, the rural exodus is due to two factors. Firstly, "*a phenomenon of rejection that pushes peasants to leave, and secondly the city, which attracts them for many reasons*" (PAULET J.P., 2006).

In Africa, as in Senegal, the arrival of settlers on the continent led to the development of areas characterised by a high concentration of people and services. These areas were initially located on the coastal fringes, as they served as transit zones and provided trade links from one continent to another. For the interior of the country, it was necessary to set up a facility so that it could concentrate[2] people in its turn in order to benefit from the opportunities offered by this infrastructure. Some people describe this period as the start of African urbanisation, while others place the beginnings of African urbanisation at the time of the trans-Saharan trade. But the realities remain the same, because African cities have developed "*As soon as a considerable investment is made, accompanied by the arrival of Western material and human resources in a neglected area, a building site is created, a temporary habitat that tends to last. This settlement can be grafted onto an existing town, which it reinforces: it can also be the beginnings of a city, with polarisation phenomena in terms of facilities and acclimatisation to the local environment*"[3] .

Over the years, these areas, often described as non-city areas, have undergone demographic, economic and spatial change. After independence, these areas continued to welcome people and are now the country's major urban centres. But this rapid urbanisation was not accompanied by advance spatial

---

[1] UNFPA, July 2007. *State of the World Population 2007. Unleashing the potential of urban growth,* 106 p.
[2] Professor Cheikh Samba Wade, 2017, Licence 1 Geography course (Urbanisation processes in Africa)
[3] Pr Cheikh Samba Wade, 2017, Course Licence 1 Geographie (Introduction to urban geography)

planning. This has led to the monstrous spatial development of some cities, with newcomers settling on the outskirts or margins of the city, as the latter can no longer contain these waves of excessive migration in its central core or in its original, historic urban core. These urban environments, inherited from colonisation as it is often said, have experienced many dynamics over time and space.
In Senegal, urbanisation was well underway before the country gained independence[4] . The country's urbanisation rate was 25% in 1960, 40.7% in 2002 and 42.7% in 2016 (ANSD, 2016). This phenomenon has accelerated, leaving the same marks on the organisation of the national territory as during the colonial period, with a concentration of towns in the west and centre-west of the country and an urban structure characterised by a metropolis that is out of proportion with the other conurbations[5] . This imbalance in the urban fabric is reflected in the difference in the urbanisation rate, with the city of Dakar at 95.5%, compared with 16% for the Diourbel region (ANSD, 2016)[6] .
But at present, policies to provide towns with infrastructure and facilities are giving rise to a new category of town, known as a "secondary town", which is becoming increasingly attractive and growing in size and number. But this urban dynamic of secondary towns poses a number of problems, including urban sprawl and its consequences. This led us to study the case of the town of Bambey.
Bambey, a secondary town that has experienced rapid demographic growth, faces a number of constraints in its spatial development. In 1988, the town had a population of 16,268, rising to 20,366 in 2002 and 28,908 in 2006.
2013. However, according to ANSD projections, the population is expected to reach 42,077 by 2025. This strong population growth means that space requirements are becoming a real problem in order to contain this growing human mass.
The aim of this study and research project is to analyse the spatial development process in the town of Bambey and its socio-spatial impact on the town and the surrounding countryside.

1.1 Introduction

The urban dynamics of towns and cities in Africa are becoming increasingly worrying these days, raising a host of questions. Today, with the growth of both primary and secondary towns accentuated by population movements and sustained natural demographic growth, cities are continuing to expand year after year, and the need for land for housing is increasing from one year to the next, and from one decade to the next.
In Senegal, despite the primacy of cities such as Dakar and the regional capitals over the other groups, the secondary cities, which are generally departmental capitals, are constantly growing in size and spreading out towards their peripheries, eating up the surrounding agricultural land. This situation is the result of the concentration of economic, political and administrative activities in these urban centres, on the one hand, and significant demographic growth exacerbated by migration, on the other.
According to MBOW D. (2014), such situations result in an increase in "the *need for housing, infrastructure and various facilities, while the population's ability to settle is somewhat limited*". In second-tier towns, these phenomena are due to the small size of the areas or initial municipal boundaries that these towns occupy, but also to the important functions that these urban centres play from an economic, administrative and even political point of view. This is the context in which the town of Bambey finds itself.
The capital of the department of the same name, the town of Bambey now receives more than 90% of government investment across the board. This has strengthened its influence over the other communes in the department, and is increasingly attracting rural populations from the surrounding countryside. In fact, the concentration of infrastructure and facilities in the department's capital has led to a real economic imbalance between Bambey and the rest of the department, in turn reinforcing the phenomenon of rural exodus.
This significant growth in the town accelerated after 2000. In fact, the establishment of Alioune Diop

[4] Lat Soucabe Mbow, Les politiques urbaines : gestion et amenagement, 1992 in codesria
[5] Lat Soucabe Mbow, Les politiques urbaines : gestion et amenagement, 1992 in codesria
[6] ANSD Report, Ed. 2016, Economic and Social Situation of Senegal in 2016

University in Bambey in 2004 was a real factor in urbanisation, "*resulting in the consumption of space and the need for building land, or at least land that can be developed in an urban setting*" (FAYE M., 2013). This urban dynamic, which always rhymes with the need for housing space, more often than not leads to tensions over land in the city.
Today, the spatial expansion of Bambey is giving rise to numerous land issues between the various stakeholders. This applies both to the authorities in the town of Bambey, who need to find land on which to settle their population, and to the surrounding local authorities, who need to protect their land on the outskirts of the town of Bambey. But these land issues go beyond that, because they also directly involve the farmers, who are in a precarious situation and want to do whatever it takes to protect their fields, which they have acquired under customary law using the system of inheritance.

1.2 Background and justification

1.2.1 Context

The world is now living in the urban era. This dynamic is not about to stop. In 2003, 53% of the population lived in rural areas, compared with 47% in cities (DIOP D. and TIMERA A. S., 2018). Thus, over the years, urban areas have undergone considerable change, both demographically and spatially. This evolution or dynamic of cities depends both on exogenous factors, the importance of which is reflected in migratory movements, and on endogenous factors resulting from residential mobility, including progressive changes in the social occupation of the city core.
In 1875, the only city in the Third World with more than a million inhabitants was Beijing, but by the 1980s, more than 150 cities were millionaires (PAULET, J. P., 2006). The urban growth of cities like Cairo in Africa is spectacular. Today, we are witnessing major growth in cities around the world, particularly in Africa and Senegal. This development of urban areas is the fruit of a long process resulting from the effects of the movement of rural populations to the cities or the movement of the affluent class from the urban centre to the outskirts or margins of the city.
In Africa, the growth of the urban population is the most important aspect of urbanisation dynamics. According to the OECD report (2020), the urban population rose from 27 million in 1950 to 567 million in 2015, an increase of 2000%. From Abidjan to Accra, via Dakar to Bamako, cities are growing in demographic size, creating a need for housing and jobs. From colonisation to independence, the dynamics of urban areas vary from one period to another, depending on the players involved.
In countries such as Cote d'Ivoire, Togo and other West African countries, cities have seen remarkable growth in recent decades. In Senegal, for example, there are more 3.1 million people living in the capital than there were in the whole country half a century ago.
Historically, urbanisation has been an economic and public policy reality in Senegal (World Bank, 2020). At the time of its accession to international sovereignty, Senegal was one of the most urbanised countries in West Africa, ahead of countries such as Côte d'Ivoire and even Ghana. In Senegal, the urbanisation phenomenon is most pronounced in the city of Dakar, which accounts for 25% of the country's population (World Bank, 2020). According to DIOP D. and TIMERA A. S. (2018), "the *hypertrophy of Dakar, the capital of Senegal, is the result of planning choices and strategies that have been mostly disjointed since the colonial period*". This justifies the imbalance in the country's urban fabric. The classification of the country's cities according to demographic size (Zipf's rank-size law) shows Dakar in first place. Followed by cities such as Touba and Thies. But at present, the development of secondary cities within the country is giving rise to the emergence of new cities linked to their regions and to the surrounding countryside, which in turn is shaping the development of the national territory.

1.2.2 Justification

The rapid growth of secondary towns in Senegal is having a number of consequences, especially in terms of space. This rapid urbanisation raises issues such as urban management to control urban development and good urban land management. The decline in agricultural yields in recent years due to rainfall variations and soil impoverishment has led to a deterioration in the rural economy and exacerbated the phenomenon of rural exodus, with more and more people leaving the countryside. The cities are the points of convergence because of the many opportunities they offer. These waves of migrants, combined with population growth in the city, are putting pressure on peri-urban areas, most of which are farmland used by rural dwellers.

The State's desire to correct the imbalance in the country's urban fabric is evident in the fact that most of the country's towns have urban planning or development documents. The creation of secondary development hubs has also meant that the country's secondary towns are increasingly benefiting from state funding for public policies such as PACASEN, PROMOVILLES, etc.

These actions taken by the central authority to develop the secondary towns mean that the urbanisation trends marked by "*a high concentration of populations in the regional capitals*" (BA A., 2015) are beginning to be reversed. In fact, some of the country's secondary towns, by virtue of their economic and administrative functions, or their role in education and research, are becoming areas for rural populations. The town of Bambey is now one of these towns that is constantly attracting people because of the opportunities it offers, especially in terms of training and research.

Bambey is a town that owes its influence to its facilities and infrastructure. As a result, it is growing steadily, eating up more and more of the agricultural land of the surrounding local authorities. This urban sprawl is fuelled by the mass exodus of people from the surrounding countryside, creating new housing and employment needs. These waves of migration are already finding the urban centre densely populated due to the "narrowness of the communal perimeter that the town occupies, a rectangle of 2,300m by 1,170m (269,10ha) means that at present, all the extensions to the town of Bambey are mainly on agricultural land in the communes of Ngoye (to the south), Ngogom (to the north), Ndangalma to the west and Thiakhar (to the east), which poses an administrative problem between these different administrative entities (PDU of the town of Bambey).

According to the Urban Master Plan for the town of Bambey, between 2001 and 2021, the population of the town of Bambey is expected to double to 53,956. This gives us an additional population of 26,982 between 2001 and 2021, which will require additional land to be developed.

The demographic pressure on the town of Bambey is due firstly to population growth, and secondly to the construction of facilities capable of accommodating large numbers of people.

On the basis of this constant, we can see that the choice of our subject of study is not fortuitous, because studying the urban dynamics of a town like Bambey and the land issues linked to these dynamics will enable us to understand :

The different aspects of urban dynamics in the town of Bambey;

Land issues related to the city's spatial dynamics ;

The impact of urban sprawl on periurban land.

1.3 Problems

The beginning of the 21^{e} century was marked by explosive urbanisation, with cities increasing in size and number. Since then, the issue of urbanisation has become a central one for all societies, in both North and South, because urbanisation has become a general, universal fact in the world[7] .

Today, more than half the world's population lives in cities. By contrast, in the 1900s, the number of rural dwellers in the world was greater than the number of urban dwellers. But by 2008, the trend had reversed, with more than 50% of the world's population now living in cities[8] . This phenomenon has a significant influence on the distribution of populations, with the resulting consequences such as a lack of access to basic social services.

In the countries of the South, the urbanisation process has been rapid, brutal and unorganised. It has been fuelled above all by waves of migration from the countryside and, according to CHALEARD J. L. and DUBRESSON A. (1999), has been accentuated by poverty in the countryside due to the

[7] Jacques Veron, Enjeux economiques, sociaux et environnementaux de l'urbanisation du monde, 2008. in carn.info, n°142. pp. 39 - 52.

[8] UNFPA, July 2007. *State of the World Population 2007. Unleashing the potential of urban growth,* 106 p.

difficulty of "*modernising low-productivity farming systems because of a lack of resources and sufficient remuneration for their work, and the rural population is doomed to exodus*".
The rate of urbanisation in Senegal has risen sharply since 1960. Between 1960 and 2018, the proportion of the population living in cities rose from 25% to 45%. In 2018, 46.7% of the country's population lived in urban areas, whereas in 2002, only 40.7% of the population lived in cities (ANSD, 2020).
In Senegal, as in most African countries, population growth in cities is a major issue. In Senegal today, the issue of dynamic urbanisation affects almost all of the country's towns, such as Bambey.
Created from small hamlets in 1886 (FAYE S., 2018), the town of Bambey has undergone various periods of development marked by different factors. From its birth to the present day, successive waves of migration continue to come and settle in the commune. These migratory movements towards the town and the significant demographic growth mean that the town has undergone a remarkable evolution over time from a demographic and spatial point of view.
Firstly, the drilling of wells in 1906 led to the first waves of migration of the surrounding populations, who settled in the current communal perimeter (FAYE S., 2018). The erection of Bambey as a sub-divisional capital in 1907 and the opening of the road station in 1908 (FAYE S., 2018) accentuated the urban expansion of the town. This urban sprawl has followed a long process since the birth of the town, with different stages and rhythms. Firstly, during the colonial era, the city's urban sprawl was a response to economic and mobility needs.
Secondly, the town's geographical location meant that it was a major crossroads between the regions of Diourbel to the east, Fatick to the south and Thies and Dakar to the west. The marketing of groundnuts at this time also played an important role in the spatial, demographic and economic development of the town.
In 1926, the town was given the status of a first-degree mixed municipality. The first subdivision plan for Bambey was drawn up in 1929 in the central core (Escale district) and in the traditional village. Thus, in 1929, the town began to develop with a perimeter of 143 ha (FAYE. S, 2018).
In 2003, Bambey benefited from a decree approving its master town planning scheme and making it enforceable[9] . However, given the town's demographic size and the narrowness of its communal boundaries, the difficulties of expanding and the new land requirements in terms of housing and facilities, the town benefited from a presidential decree in 2011 which extended its communal boundaries[10] .
Since its creation or birth, the town of Bambey has continued to grow, but this growth has been greater and faster from 2000 to the present day (2020). The establishment of the Alioune Diop University in Bambey (2004) and the installation of modern sports and health facilities have made the town more attractive, and its level of influence is now greater than ever, even surpassing the other regions of Senegal.
Added to this are the natural constraints of the town of Bambey. Indeed, the town's main problem is the systematic flooding of its entire municipal territory during the rainy season. This flooding is partly due to the fact that the town is located in an area of depression that used to be home to a network of marshes that drained all the rainwater from the town, and partly due to the nature of the soil, which is clayey and not very permeable, making it difficult for run-off water to infiltrate. As a result, some people have moved from the town centre to the outskirts, where they are less affected by flooding during the winter months.
Today, because of the exhaustion of the town's communal perimeter and its need for land, facilities and housing, the Bambey authorities have initiated a subdivision project for 2020. The areas to be subdivided as part of this project, apart from being agricultural land, traditionally belong to the

---

[9] Decree no. 2003-410 of 4 June 2003 approving and making executory the urban development master plan for the municipality of Bambey
[10] Decree no. 2011-1991 of 16 December 2011 amending and extending the territorial boundaries of the municipality of Bambey in the Diourbel region

surrounding communes such as Ngoye to the south, Thiakhar to the east, Ngogom to the north and Ndangalma to the west. This subdivision plan is the subject of several debates between the local authorities concerned and the farming population. This highlights all the constraints on the spatial expansion of the town of Bambey.

1.3.1 Research questions

Main question

In this work, the main question is:

What are the different aspects of urban dynamics in the town of Bambey?

Specific questions

1. What are the different factors in the spatial dynamics of the town of Bambey?
2. What are the land issues related to the urban dynamics of the town of Bambey?
3. What is the impact of urban sprawl in the town of Bambey on periurban land?

1.3.2 Research hypotheses

These research questions give rise to the following research hypotheses:

Main hypothesis

Bambey is a secondary town located at a crossroads in the western centre of the groundnut basin, and has a number of features that accentuate its urbanisation dynamic.

Specific assumptions

1. Rapid population growth combined with the development of infrastructure and facilities are the main factors behind Bambey's urban dynamics.
2. The spatial dynamics of the town of Bambey are at the root of the many issues surrounding land.
3. The urban sprawl of the city has had numerous impacts on periurban land.

1.3.3 Research objective

***J* General objective**

The general aim of this study is to analyse the urban dynamics of the town of Bambey, and the land issues associated with this urban phenomenon.

Specific objectives

In addition to the main objective of this work, the specific objectives are to :

1. To study the main factors in the spatial development of the town of Bambey.
2. Identifying the various land issues arising from the spatial expansion of the town of Bambey.
3. Analysing the impact of urban sprawl in the town of Bambey on land in the surrounding communes.

1.4 Conceptual clarification

As in all research work, it is always essential to clarify the key concepts in order to remove any ambiguities, gain a better understanding of the subject and define our angle of approach in this work. The key terms in our subject were defined on the basis of our reading.

In this work, the words or groups of words to be clarified include urban dynamics, urbanisation and the concept of the city and land.

Urban dynamics

Quite simply, the concept of dynamics refers to an evolution, something that changes over time and, in the case of certain phenomena, most often in space. But the adjective "urban" attached to the word "dynamic" gives rise to the desire to define the word "dynamic" succinctly in relation to its qualifier. According to BRUNET R., et al (2005), the concept of dynamics can be defined as: "*A change resulting from an interplay of forces;* [...]". On the other hand, according to Jean Romeo, dynamics can be qualified as a theory when it aims to interpret a situation that results from influences produced at different times and manifested at the same moment or generated at the same date and felt successively[11] . From this point of view, urban dynamics have a temporal dimension and can be

[11] Jean Romeuf, quoted by Aminata, nee Caroline Ba in her master's thesis entitled, *Dynamique d'urbanisation de la ville de Thies,* 2015

analysed in relation to the expansion of the city, the management of urban land and also the management of the city's facilities.

To study the dynamics of an area, dynamic models relate changes to the forces that cause them. Urban dynamics, like territorial dynamics, studies changes in the organisation of a territory and the forces that generate them. The dynamics of a territory or an area can be read and analysed in terms of the location of activities, equipment, overall population and categories of people (BRUNET R., et al., 2005). To study the dynamics of a city, maps and time-varying satellite images are a good way of expressing these dynamics. But the dynamics of a space must not be confused with everything that is mobile in space, such as the movement and/or mobility of goods and people (BRUNET R., et al., 2005).

Urban dynamics, if considered as the development or growth of cities, has three aspects:

The first is spatial: here we need to know how, step by step, the site has been extended;

The second is functional, and here the question is what succession of functions and adaptations to historical circumstances has led to the city we know today;

J The third is the demographic aspect (DERRUAU M., 2002).

In fact, the function sets a population for which the town must multiply its buildings. As with any population, the city's population varies through natural growth and migration. The percentage increase in population between two censuses is a good indicator of a town's dynamism.

In our work, urban dynamics refers to the process of city development. It refers here to patterns of land use, the distribution and growth of basic social services, and the different stages in the development of urban space over time and space.

To speak of the urban dynamics of the town of Bambey is to define the term as the process through which the town undergoes its spatial development. However, it is more judicious to include the different aspects because this dynamic is subject to events that take place in space over time. These facts can be the construction of infrastructures and facilities in the city which attract more people and accentuate the demographic growth of the city. The urban dynamic of Bambey can be measured by the expansion of the town, the growth of the built-up area and the reduction in agricultural land, as the urbanisation of the town has taken precedence over the development of agriculture. This paradox shows the socio-economic conversion of a town that once played an important role in the groundnut trade, especially during the colonial era. This is to say that, over time, the town's main functions can be reversed in favour of other activities. Today, in addition to its main function as a port of call for the groundnut trade, Bambey has political and administrative functions, and since the 2010s has been regarded as a training and research town, a university town.

Urbanisation

According to the dictionnaire de l'urbanisme et de l'amenagement, *"the term urbanisation has two distinct meanings"*. Firstly, it is understood as "the action of urbanising, creating towns or extending urban space". Secondly, and more frequently, it refers to the growing concentration of populations in cities. But for BRUNET R., et al (2005), in the critical dictionary (les mots de la geographie), urbanisation is the development and expansion of the urban population, and this current urban phenomenon is taking place outside the cities, on the outskirts of the city.

It is a process, whether controlled or undergone, characterised by the growth of cities, which spread out to their peripheries and nibble away at rural areas. It is a universal fact, noted throughout the world and manifested by the extension of the surface area of the city resulting from the increase in population and functions existing in the city studied (MBOW L.S., 1992). This phenomenon occurs rapidly in developing countries. In these countries, the urban population had exceeded 50% everywhere by the end of the 19^{e} century, and stood at 75% in England (MERLIN P. and CHOAY F., 1998). In Third World countries, however, this phenomenon has accelerated since the 1950s. Between 1800 and the middle of the 20^{e} century, the proportion of city-dwellers in these countries rose from 8% to 15%, before reaching 28.4% in 1980 (MERLIN P. and CHOAY F., 1998).

The urbanisation of a locality can be linked to the political function it performs. But in very general

terms, according to the dictionary of history, economics, finance and geography, urbanisation also refers to all the changes in social habits and material living conditions (housing, consumption, travel and work) associated with the presence of cities (TEULON F., 2010).
To study and understand the level of urbanisation of a continent, a country, a region or a city, it is necessary to measure the ratio between the number of urban residents and the number of rural residents, to assess the density of different areas and the spatial extension of cities, and also to evaluate the transformation of lifestyles.
In the town of Bambey, assessing the lifestyles or economic activities of the town's population is tantamount to saying, in principle, that we are in the middle of a rural environment, since the lifestyles and behaviour of the population are purely and simply rural practices. However, in the context of our work, we are not very interested in lifestyles or consumption patterns, and even less in the social practices of the population. On the other hand, here we are talking about the town of Bambey as a living space, a place to live, the physical support and its constructions, even if the place of certain practices cannot be excluded in this study. This commune is characterised by the development of buildings in all its neighbourhoods. Defining the urbanisation of Bambey is tantamount to talking about the development of the built environment and its many consequences for land. Whatever its lifestyles, the physical or geographical urbanisation of Bambey no longer needs to be proven, and indeed it far exceeds the sociological urbanisation of the town's population.
In the context of our study, the concept of urbanisation takes on its full meaning from the point of view of the dynamics experienced by the city in a spatial approach. In this work, it is the physical aspect of the urban phenomenon that seems to be more important than the other aspects highlighted by urbanisation.
-S **Property**
From the Latin "fundus", which means "bottom of the earth", "foncier" in its original meaning refers to land and sometimes, by extension, to the buildings and facilities on it, not for its own sake, but as the basis of a farm or an income. It is a multidisciplinary concept and its definition varies from one discipline to another. In economics, it is the idea of rent, of value, that will be put forward to define land, whereas in a much more sociological approach, it will be a question of demonstrating how the land variable plays a part in the efficiency of the agrarian system, for example[12] .
Legally, the term "foncier" *refers to "property that is immovable by nature, i.e. that does not circulate, so the soil and those who live on it are fundamental attributes of foncier"* (CASSE L. O., 2010). In other words, land refers to land that does not circulate. So the soil and its fixity are fundamental elements of land tenure. In a global and interdisciplinary approach (legal, economic, geographical and sociological), the term "land" becomes a total social fact that sets in motion all the components of society and reveals all its facets (LE BRIS E. et al., 1982). In the same vein, the Petit Larousse defines land as immovable property. It also refers to building land.
In geography, land tenure refers to a set of relationships involving the organisation of space (LE ROY, 1982)[13] . This concept implies the idea of appropriation, transfer, occupation and management of land. But according to the FAO, it refers to "*a set of relationships between individuals and the land and the renewable resources it supports*"[14] .
In our work, talking about land means discussing this tradable good, this good that supports production and consumption activities (agriculture and livestock farming). This means highlighting the aspect of land use, whatever the purpose. Over and above the legal aspect of the term, the economic and geographical aspect is of great interest to us here.
In the context of this study, the land we are talking about refers to the vast expanses of agricultural land that have been owned and farmed for decades by farming families and which are passed on from

---

[12] E. Le Bris, 1982 quoted by OUATTARA S. 2007 in his doctoral thesis in sociology
[13] Le Roy E., 1982. Enjeux fonciers en Afrique Noire, cited by CASSE L. O. 2010 in his master's thesis in geography.
[14] See FAO definition cited by Sencebe Y. et al., 2013. "TERAGI/Volet foncier : problematique et methodologie", in journal open edition, n°3, vol. 4, consulted on 30 November 2022

one generation to the next by the process of inheritance in accordance with the principles and operation of customary law, which, moreover, gives recognition to the owners. For the purposes of this work, the land we are talking about is the physical environment, the soil, which supports economic activities such as agriculture and livestock farming. Today, however, with the advance of the built environment, it has become a commercial asset that is developing a vast uncontrolled market, leading to total land insecurity and, in turn, numerous problems between administrative districts or between social groups. As in all areas where rurality dominates over urbanity and economic activity is essentially based on the primary sector, the relationship between social groups and land is much stronger. The idea that a plot of land belongs to a particular family is still alive and well in the town of Bambey and the surrounding area. But today, whatever the case, land remains the physical support of the living environment with the buildings that stand on it.
According to FAYE J., land tenure is defined by specialists as all the relationships between men and women concerning land and the resources found on it. In most African societies, obtaining an area of land for agriculture and housing is the main concern of social groups[15] .
In Senegal, the land issue is still very topical today, as it is directly linked to the dynamics of the cities and gives rise to disputes between social groups.
As part of our study, we looked at this concept as a medium for economic activity in the past, but which has now become a commercial asset with a growing market.

***J* Ville**

First of all, we need to point out the vague and imprecise nature of the term city. The nation of the city seems perfectly clear to all of us, but we stumble as soon as we try to give a definition (DERRUAU M., 2002). This is perhaps what led Jacqueline Beaujeu-Garnier (1997) to emphasise the difficulty that arises when we try to define the notion of the city: "*I know what the city is as long as I am not asked to define it*". However, there are a few criteria that can be used to define the city.
According to the dictionnaire de l'urbanisme et de l'amenagement (1992), three conditions are essential for a human settlement to constitute a town: a building agglomeration, a large cluster of contiguous houses, a building meeting, certain features of the population and status in the 18^{e} century. Based on this logic, Bambey is a town. Thus, (MERLIN P. CHOAY F., 1992) assert that the city is fundamentally born of the central functions of exchange. On the other hand, the Larousse dictionary defines a town as a relatively large agglomeration whose inhabitants have diversified professional activities. Le Robert, on the other hand, defines a city as a group of buildings whose inhabitants work in commerce, industry or administration. This definition, based on the activities of the population, gives Bambey the status of a country town. In fact, according to this definition in the Robert dictionary, the town of Bambey cannot be considered a town because most of its inhabitants work in agriculture and livestock farming. For DERRUAU M., in geographie humaine, 2002, the city is a durable agglomeration as opposed to temporary agglomerations.
But when it comes to defining the city, as soon as other criteria, such as the statistical criterion, are considered, difficulties arise, because there is a multitude of cities from one country to another, depending on the demographic size of the urban areas. But despite everything, some countries define the city on the basis of the number of inhabitants, while others combine the number of inhabitants with the sector of economic activity.
In France, a town is an urban unit made up of one or more municipalities with a population of over 2,000 and a continuous population (Dictionnaire de l'Urbanisme et de l'Amenagement). According to DERRUAU M. (2002), until 1999, a town in France was defined as "*any agglomeration of more than 2,000 inhabitants clustered around the main town of the municipality*". The limit is 5,000 inhabitants in Austria and 300 in Iceland (MERLIN P. and CHOAY F., 1992).
However, this statistical criterion poses a problem everywhere, as there are many cases of large

[15] Faye J., 2008. "Land tenure and decentralisation. L'expérience du Senegal, in foncier-developpement.fr, consulted in December 2022.

farming communities that are not towns, such as the villages of 10,000 inhabitants in the Tonkinese delta or the countryside of 50,000 inhabitants in China. Others look for a definition in the activities of the inhabitants at a time when some combine demographic size and non-agricultural activities. But a definition of the city in terms of non-agricultural activities comes up against a new objection. There are some very large towns in the world known as agrovilles, agglomerations of farmers with the formal appearance of towns (DERRUAU M., 2002). The town of Bambey can be classed as one of these agri-towns, even though agricultural activity is beginning to be overtaken by other economic sectors such as commerce and real estate.

According to DERRUAU M., a town can therefore be defined as "*a large agglomeration laid out for collective life (this layout constitutes town planning) and in which a significant part of the population lives from non-agricultural or agricultural activities*". It is therefore impossible to set lower limits to define a town, because there are so many different types of town, and it is necessary to study both their diversity and their common character (DERRUAU M., 2002).

In Senegal, the statistical character is not so decisive in the definition of the city because the latter refers to the commune. Thus, every commune is a town if we are to believe the general code of territorial collectives, which states that: "*The town has the status of a commune*". It results from the pooling of the competences of several municipalities that are territorially homogeneous [16] Thus, the urban environment is made up of all the localities designated as communes, regardless of the number of inhabitants. Thus, the urban population is that which resides in the communes (ANSD, 2013). Based on this definition, Bambey would not be considered a town. In fact, the town of Bambey is not made up of several communes, because in terms of territorial management, the commune of Bambey has no powers over the other communes in the department, unless we are talking about a departmental council, or if this is the case, it is in the context of strictly administrative functions.

So it is clear that to speak of Bambey as a town is to say that it is a place where the built environment and the development of economic activities are very important compared with the other communes with which Bambey shares the department of the same name. Bambey ville is a place where markets operate on a daily basis, despite the existence of weekly markets. From an administrative point of view, it is a place that houses many administrative buildings and therefore plays a very important political and administrative role in the department's administrative architecture. So, when we talk about the town of Bambey, we are referring to this commune which has command powers over the other communes in the department which surround it or form a whole set of relationships whose central nucleus is the town of Bambey.

I.5 Research methodology

1.5.1 Documentary summary

From the outset of this work, we focused on a documentary search based on the themes of the city, urbanisation, urban sprawl and land governance in Senegal by looking through laws, decrees and manuals of procedures on the process of land allocation by communes. During this documentation phase, we went to the central library of the Gaston Berger University in Saint-Louis and to the Documentation Centre of the UFR de Lettres et Sciences Humaines of the same university.

As part of our research, we also found online libraries very useful. We visited online sites such as that of the ANSD, the digital library of Cheikh Anta Diop University and the online library (Scholarvox Senegal). Some of these sites, such as that of the ANSD, enabled us to see the annual reports on the economic, social and regional situation of the Diourbel region, and of Bambey in particular.

For the document review, we have chosen to produce a summary by theme and by the objectives that were pursued when reading each reference, rather than a document-by-document summary.

We visited Bambey town hall and the Bambey departmental urban planning and housing department. This enabled us to consult the Reference Urban Development Plan (PUR) for the town of Bambey, as well as its Urban Development Master Plan (PDU), to gain a better understanding of the town's

[16] Law no. 2013-10 of 28 December 2013 on the General Code of Local Authorities

functions and spatial organisation.
Our documentation phase enabled us to consult a large number of references focusing on the themes of cities and urbanisation. These included the book **Introduction a la geographie humaine** (BAILLY A. and BEGUIN H., 2016). This book gave us a general idea of the evolution of the urban phenomenon in the world. It is similar to DERRUAU M.'s work, entitled **Geographie Humaine** (2002) which, in addition to the conceptual elucidations it offered, gave us a view of the state of the world's urbanisation dynamics and the production of urban spaces. He also discusses the methods that can be used to study this exponential urban growth around the world, which has been much more significant in the countries of the South since the 2000s. In this book, Max DERRUAU also describes the various aspects that enable us to appreciate very clearly the dynamics of urban areas. Another reference work is BAILLY A. et al (1998) entitled **Les concepts de la geographie humaine**. In a way, this book explains the different concepts used in human geography. This conceptual clarification is sometimes accompanied by a few examples which have also enabled us, for example, to make the difference between what is a city and what is not, given that not every group of people is a city.
Reading the book **Villes et campagnes dans les pays du Sud** (CHALEARD J.L., and DUBRESSION A., 1999) gave us a focus on the causes of the dynamics of urban areas, whose dynamism is manifested by spatial extensions of the city, the multiplication of functions in the urban centre, characterised by multiple diversity as new functions are added to the original ones. Thus, in this book, we understand perfectly well that the determinants of city growth or of the dynamics of urban spaces are not obviously rooted in the space of the city, but are in fact exogenous. This urban dynamic can be explained by the crisis in the countryside, with all its corollaries. This leads us to distinguish between endogenous and exogenous causes of the urbanisation dynamic.
Books such as **Geographie Urbaine** by PAULET J. P. (2006) and **Geographie Urbaine** by BEAUJEU-GARNIER J. (1997) deal with issues relating to the city and all its phenomena. According to these two works, these phenomena include urban growth and the crisis of urban spaces, among others. In **Geographie urbaine,** PAULET J. P. (2006), questions relating to the evolution of urbanisation in today's world occupy an extremely important place, as it relates very explicitly the revolution and the distribution of the population in the urban world compared with that in rural areas over time. Like the work by PAULET J. P. cited above, that by Jacqueline Beaujeu-Garnier also gives us an overview of the development of the urban phenomenon throughout the world. By reading these two works, we have come to understand the essential factors that determine the evolution of cities in size and number throughout the world, in time and space.
The books ***Amenagement du Territoire: Principes, pratiques et devoirs pour le XIXe siecle*** by M. M. DIAKHATE M. M. (2011) and ***Le Senegal, quelles evolutions territoriales?*** by MANGA C. T. et al (2012) have enabled us to understand the territorial organisation of Senegal. This territorial organisation involves seeing and understanding the distribution of people and activities in the national territory. In this way, the territory is shaped according to the arrangements of time. Thanks to the work of (DIAKHATE M.M., 2011), we understand that the imbalance of the country's urban framework is the consequence of the choices made by the authorities of the time in implementing a policy or a land use plan.
The work by (MANGA C. T., 2012) focuses on the evolution of Senegal's territory and the factors behind this territorial dynamic. Thus, the perspectives and strategies to be adopted for a better organisation and balance of the territories were the object of a careful look within the framework of our research. It deals with issues relating to the transformation of urban areas in Senegal. These spatial changes are being driven by the various public policies of the State. According to MANGA C. T., the territorial organisation of Senegal is subject to the public policies of the colonist and the various governments since our accession to independence. In this book, too, the importance of decentralisation in shaping our territory occupies an important place and has enabled us to understand how planning policy choices influence the organisation of spaces.
The reference, ***La recomposition des espaces urbains et periurbain face aux changements***

climatiques en Afrique de l'Ouest, edited by MBAYE I. in 2018, tells us about the reorganisation of spaces, the mutation of spaces and their current form in the country and what they were in the past. The work shows that many factors are involved in these spatial changes. These include climatic, social, demographic, economic and political factors. But the book also highlights the direct and indirect impact of climate change on spatial change in urban and periurban environments.

Thanks to the work of **LE BRIS E. et al** (1982) entitled "***Enjeux fonciers en Afrique Noire***" (***Land issues in Black Africa),*** we have understood the evolution of land governance and land legislation in Black African countries. This document, which is a compilation of debates and workshops, looks at the issues surrounding land tenure in cities and the methods of securing land tenure adopted by the different powers at different times. The work of FALL A. S. et al. ***Urbain-rural : l'hybridation en marche"*** (2005) has enabled us to understand the development of our cities and the different phases involved. The example of the city of Dakar, which began to develop in 1973 with the agricultural crisis and the emptying of the countryside, is a good illustration of the relationship between the countryside and the city.

This shows the invaluable contribution of general and specific works to this study and research work. However, with dissertations, theses and scientific articles, the research becomes very interesting because these references deal with a specific subject for a well-defined area and time.

But it is also important to emphasise the vital contribution made by study reports on issues such as cities and development, which have greatly helped us to better define the angle of analysis for our subject. These reports include those by the ANSD (2013), the World Bank (2015) entitled "*Villes emergentes pour un Senegal emergent*" (*Emerging cities for an emerging Senegal)* and the OECD (2020) entitled "*Dynamique d'urbanisation de lAfique*" *(Urbanisation dynamics in Africa)*. The ANSD 2013 report provided us with statistical data on urbanisation in Senegal and the population of the country's cities, as well as qualitative data that enabled us to understand the country's urban fabric and the new trends that are emerging with the rise of certain secondary cities such as Bambey. The World Bank's report gave us an insight into the evolution of the urban phenomenon around the world and the characteristics of the phenomenon in Third World countries, particularly Senegal, which has been marked in recent years by the spatial and demographic development of cities. As for the OECD 2020 report, it deals with issues specific to the dynamics of African urbanisation in 2020, marked by urban sprawl, the in situ urbanisation of rural areas, the formation of metropolitan cities, the level of urbanisation, the pace of urban growth and the growth in the population of cities, as well as the constraints on the spatial extension of cities and urbanisation that goes beyond administrative boundaries. The 2011 UN-Habitat report entitled **"*Housing the poor in African cities*"** also helped us to understand the problem of access to land in cities and the difficulties encountered by low-income populations in finding housing.

It should also be noted that some work has been done on the theme of our research, including the work of (BA, A., 2013). However, we are not aware of many works relating to this theme as it applies to our study area. However, we can cite the dissertations submitted on the city of Bambey, but before that, it is also important to go back to the doctoral thesis by DIOP A. (2005) entitled: "***Ville et amenagement du territoire au Senegal***". This is a very important reference for our work. In fact, this thesis explains the birth of towns and communes in Senegal, as well as their spatial development process, which has different aspects from one town to another or from one commune to another. The various factors in the growth of towns and cities are therefore an important part of the work. In his master's thesis in geography, (FAYE S., 2018) entitled ***"Dynamique spatiale et les enjeux de I'amenagement du territoire : cas de la ville de Bambey" (Spatial dynamics and land-use planning issues: the case of the town of Bambey)***, he discusses the town's spatial dynamics and land-use planning prospects. However, the inadequacy of this work is underlined by the fact that the contribution of cartography and remote sensing is too weak in relation to the subject, since talking about the spatial dynamics of a town means using maps or satellite images to provide a better illustration of key periods defined in advance. In fact, studying urban dynamics amounts to a diachronic analysis of the environment over a

well-defined period of time. So we understand the importance of cartography and remote sensing tools. This is why cartography and remote sensing will play a key role in our work. The work of (FAYE M., 2013), ***"L'implantation de I'Universite Alioune Diop de Bambey comme facteur d'urbanisation"*** (***The establishment of the Alioune Diop University of Bambey as a factor in urbanisation***), has enabled us to learn about the involvement of the Alioune Diop University of Bambey in the rapid development of urbanisation in the town. However, we have to admit that, in our view, the work presents a number of points that needed further clarification, such as the diachronic analysis of the space before and after the establishment of the university. Nevertheless, thanks to this work, we have understood the role and impact of the university in the development of the city. For the work of (Faye A., 2013)
"Migration internationale et urbanisation: I'apport des emigres dans le développement de la ville de Bambey", master's thesis in geography, Cheikh Anta Diop University, Dakar, it can be said that the importance of this work lies in the clarification it provides on the urban development of the town of Bambey, even though the main problem with the other works is the lack of illustration using cartographic tools. However, we all understand the importance and place of GIS in the study of the spatial evolution of towns. All this research has given rise to in-depth reflection and updated information on the issues surrounding the dynamics of urbanisation in the town of Bambey. These dynamics are both spatial and demographic. In addition, these studies have revealed the role and place of the functions performed by the town over the course of history in the current dynamics.
In addition, the city's town planning documents, such as the **Master Town Planning Scheme** (PDU) and the **Reference Town Planning Scheme** (PUR), have enabled us to understand the spatial organisation of the city of Bambey, and give us a clear picture of the level of infrastructure and facilities in the city. Thanks to the PDU, for example, we have understood that the town of Bambey has exceeded its territorial limits and is in a situation of land scarcity.
ANSD reports such as the general report on the population census of 1978, 1988, 2002 and 2013 have enabled us to see how the population of the town of Bambey has changed since 1978. With the support of the annual regional reports for the Diourbel region, we have the rate of urbanisation of the town since 1978, 1988, 2002 and 2013, as well as the problems encountered by the population in terms of sanitation. These ANSD reports have helped us to understand the city's economic situation, which is based almost entirely on agriculture and small businesses.
In summary, it should be noted that there is not a great deal of work on the town of Bambey. However, thanks to our surveys and interviews, as well as the contribution of satellite images, we were able to achieve the desired results.

1.5.2 Surveys and interviews

1.5.2.1 Field surveys

In this Study and Research Report (SRR), we carried out fieldwork to collect first-hand information. This is quantitative data collected from households. This data was obtained as part of our survey. In this survey, the target neighbourhoods are mainly those on the outskirts of the city, as we hope that working on this part of the city will give us a better understanding of the city's urban dynamics, given that 9a has enabled us to find out what attracts people to the city until certain populations move to the outskirts. It has also given us information about their standard of living, their occupation, etc., which we feel is necessary for our research. This is what we feel is necessary to find out about the economic situation of the inhabitants of these new districts.
This survey of households in the town of Bambey was based on the number of households per neighbourhood. In other words, we took a sample of 2% of households in each district of the town. But given the number of households per district, we can also say that we surveyed 2 households out of 100 for each district.

Table 1: Total number of households per neighbourhood and number of households to be surveyed

Neighbourhoods	Number of households	Number of households to be	Percentage

		surveyed	
DVF	1234	25	2%
Leona Nord	500	10	2%
Leona South	799	16	2%
Wakhaldiam	729	15	2%
TOTAL	**3262**	**66**	**2%**

Source: ANSD, Gueye 2022

We chose the kobotolboox survey software for designing the questionnaire, administering it and collecting data in the field. This choice of software is justified by the fact that it is one that we have mastered very well, but the most important thing is that this software also allows us to produce a cartographic representation of all the households surveyed, and that the survey report offers more possibilities for producing graphs or tables to present the survey data. Another advantage is that data processing is easier with this software, because as soon as the survey is started, the software carries out the processing at the same time. This makes it easier to use the data and saves time.

1.5.2.2 Interviews

As regards interviews, we visited Bambey town hall and the Bambey departmental urban planning and housing department. This enabled us to speak with the head of the technical department at Bambey town hall, **Mr Mansour Ndiaye, on** 23 August 2022, and also with the head of the departmental urban planning and housing department in Bambey, **Mr Khadim Fall,** on 25 August 2022, as well as with a member of civil society to discuss land disputes in the town.

The aim of these interviews is to understand the role of these structures in the urban management of the town. The aim is also to understand the land issues that pit the town of Bambey against the surrounding communes. The interviews also sought to understand the authorities' plans for controlling the town's spatial growth.

These interviews were of great importance to us because they enabled us to gather vital qualitative data for our work. This strengthens our argument.

The dynamics of urbanisation have meant that the city is spreading out beyond these administrative boundaries, even nibbling away at some of the agricultural land belonging to the former rural communities of Ngoye, Thiakhar, Ndangalma and Ngogom, which became communes under Act 3.

some of the municipal councillors of these communes to get their ideas on the situation between Bambey and their communes.

The interview guide is used for these interviews.

However, it should be remembered that the interview guide is drawn up according to the resource person with whom we are interviewing. This means that we designed an interview guide for each resource person, given that the areas are not the same. However, the main focus is the town of Bambey.

1.5.3 Statistical data

To carry out a field survey, an exact knowledge of the site's statistics is essential, because in a job like this, we need to know the number of inhabitants in the town of Bambey, and the number of households in each district or in the districts targeted by our survey. To find out the population figures for the town, we visited the ANSD website to consult the 2013 population census report. This gave us the demographic data we were looking for, which we then processed to determine the size of the survey sample. For the purposes of our work, we decided to work with data from the 1972, 1988, 2002 and 2013 census reports, even though we sometimes used data from population projections for 2018, 2021 and 2022.

For data relating to requests for plots of land and housing estates that have been built or are in the pipeline, we went to the departmental urban planning and housing department in Bambey to collect the number of requests for plots of land per year, because having these statistics gave us an idea of the urban dynamics of the town and the methods used to acquire plots of land. But the department in turn referred us to the Diourbel regional town planning division. This was because the Bambey departmental urban planning department was set up in 2017, with data missing from previous years.

1.5.4 Satellite images

Even if the urbanisation dynamic of a city can be lost from a demographic point of view, this is not enough, because the growth of a city, if it is felt at all, is above all in space, with the built-up area expanding towards the periphery. This phenomenon is best observed with cartography technologies, thanks to telegestion and satellite images. In this research work, a touch of remote sensing was used to carry out a diachronic analysis of the city, giving us a better idea of its spatial evolution. This diachronic analysis of the city was carried out over 30 years starting in 1990, using a 15-year time step. The dates 1990, 2005 and 2020 are the benchmarks for this study. We will therefore analyse the land use of the city in 1990, 2005 and 2020. Three (3) land use maps of the city are produced.

To produce these maps, it is necessary to acquire satellite images. From the USGS website, we downloaded three (3) Landsat 7 satellite images for the years 1990 and 2005 and one (1) Landsat 8 image for the year 2020. The images were processed using QGIS and R software. The images were sliced and classified using R software. To classify the images, we used the NDVI of the 1990 image to highlight the vegetation, buildings and empty spaces in the image. Using this model, we classified the images (2005 and 2020) and then produced the land use maps.

We then calculated the surface area of buildings, empty spaces (bare soil) and vegetation. This gave us a clear idea of the evolution of the areas occupied by these three classes. All this basic work was carried out using R software.

The maps are produced using a combination of QGIS and R. The use of these two software packages is explained by the limitations of one or the other in image processing.

As with any scientific model, this working model was subjected to an evaluation to assess the quality of the results obtained. We used two methods to evaluate the results of our study and research.

The first is the global accuracy model. This is a model that essentially tells us the proportion that has been mapped correctly. Overall accuracy is generally expressed as a percentage, with 100% accuracy being a perfect classification where all the reference sites have been correctly classified. In the end, this model only provides the user and the map producer with basic information about accuracy.

For our work, we have an average overall precision **of 0.75%, i.e. 75%** precision.

The second method is the evaluation model for calculating the Kappa coefficient. The Kappa coefficient is generated from a statistical test to evaluate the accuracy of a classification. Kappa essentially evaluates the performance of the classification compared with the random assignment of values, i.e. whether the classification did better than randomly. The Kappa coefficient varies from -1 to 1. A value of 0 indicates that the classification is no better than a random classification. A negative number indicates that the classification is significantly worse than random. A value close to 1 indicates that the classification is significantly better than random.

According to the calculation of the "Kappa coefficient", the result of the image classification is **0.63, i.e. 63%** accuracy. But beyond these two evaluation models, other measures of precision such as the precision of the producer and user side are called upon in this work for greater scientific rigour.

Producer accuracy is therefore the accuracy of the map from the point of view of the cartographer (the producer). It is the frequency with which actual terrain features are correctly displayed on the classified map, or the probability that a certain land cover of an area on the ground is classified as such. Producer accuracy is the complement of the error of omission, producer accuracy = 100% error of omission. It is also the number of reference sites accurately classified divided by the total number of reference sites for that class.

User accuracy is the accuracy from the point of view of the map user, not the cartographer. User accuracy essentially indicates how often the class on the map will actually be present in the field. This is known as reliability. User accuracy is complementary to commission error, user accuracy = 100% commission error. User accuracy is calculated by taking the total number of correct classifications for a particular class and dividing it by the total for the line. This model is based on an error matrix. After evaluation, here are the results of the producer and user accuracies for the 2005 and 2008 images.

Table 2: Results of the accuracy of the Landsat 7 2005 image classification model

Classes	Precision of the producer	User precision
Habitat	0,78	0,81
Vegetation	0,87	0,82
Empty spaces (bare ground)	0,91	0,93

Table 3: Results of the accuracy of the Landsat 8 2020 classification model

Classes	Precision of the producer	User precision
Habitat	0,51	0,67
Vegetation	0,71	0,62
Empty spaces (bare ground)	0,80	0,66

The results of these precisions show that image classification is generally good. In fact, all the methods used to calculate the precision show that we are above 50%, which is satisfactory overall and presages good image classification.Summary table of the field data collection plan and tools

Table 4: Summary of the field data collection plan and tools

Periods	Locations	Targets	Methods	Techniques	Tools	Workforce
October 2022	City of Bambey	Populations of the different neighbourhoods	Quantitative	Maintenance Individual	Questionnaire	66
September and October 2022	Places of work or service	Head of the Technical Department of the Bambey Town Hall; Head of the Bambey Town Planning and Housing Department; - One (1) member of the non-trading company	Qualitative	Maintenance Individual	Maintenance Guide	3

I PART

URBAN DYNAMICS IN THE TOWN OF BAMBEY

In this section, we first present the town of Bambey. This means describing the physical characteristics of the soil and the climatic characteristics of the environment. Secondly, a brief overview of the town's economic activity sectors, and thirdly, a description of the town's urbanisation process, before concluding with an analysis of the various factors in the town's spatial dynamics.

Chapter I: Presentation of the town of Bambey

In this chapter, we present the town. We then talk about its geographical position, its characteristics, especially the physical characteristics of its site, i.e. the nature of the soil and also the climatic aspects of the area. However, we have also analysed the city's various business sectors and its spatial organisation.

1.1 A town at the heart of the groundnut basin

Located in the centre-west of Senegal, in the Diourbel region, Bambey is a secondary town in the middle of the countryside, providing a link between Dakar, the capital of Senegal, and its hinterland. It lies between latitude 14°42 North and longitude 16°27 West. The town is located in the department of the same name. The town lies halfway between Thies and Fatick, to which it is linked by the two inter-regional roads R.32 and R.61, and 20 km from Diourbel on Route Nationale 3 (RN3) linking Dakar to Diourbel via the town of Thies.

The town of Bambey is at the crossroads of a major road and rail network, making it a crossroads. This ease of access fosters the development of major exchanges between the town of Bambey and its hinterland, which is mainly used for agro-pastoral purposes, on the one hand, and the rest of the country on the other (Bambey PDU).

The town of Bambey lies at the centre of the groundnut basin and is bordered to the north by the commune of Ngogom, to the south by the commune of Ngoye, to the east by the commune of Thiakhar and to the west by the commune of Ndangalma. It is mainly in these four communes that the town's spatial dynamics are most apparent.

Despite its status as departmental capital, Bambey is the smallest commune in terms of surface area in the department. It covers an initial area of 7 km (^{2}), which with recent extensions has grown to 9 km (^{2}).

Map 1: Geographical location of the town of Bambey

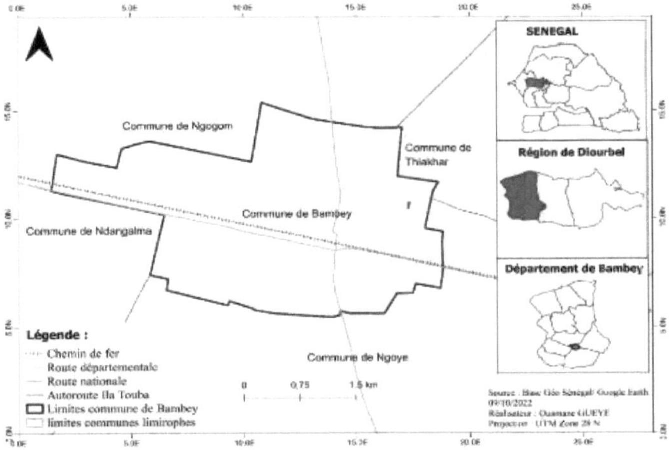

Source: Base Geo Senegal/Google Earth/QGIS

1.2 Physical and climatic characteristics of Bambey

A town with rural characteristics, Bambey's location means that it is affected by the Sahelo-Sudanian climate from the north of the country. These climatic features, which are severe for most of the year (dry season), mean that the area's climatic conditions are restrictive under certain conditions. This influences the state of the vegetation's soils and the city's water conditions.

The town's site is marked by a relatively homogenous relief, with vast sandy plains and depressions in the southern part of the town. This facilitates silvipastoral activities in the department in general and around the town in particular. The geological formations on the site belong to the Tertiary era, i.e. the "Lower Eocene" with limestone marl. The town site is an ancient depressional zone that once formed a hydrographic network with the marigot.

In most of the city's districts, some pools of water are still visible, especially during periods of heavy rainfall, when rainwater flows into the districts. The impermeability of the soils and their hydromorphic nature with low porosity cause accidents, and the intensity of run-off and the formation of overflowing pools are more noticeable.

Within the municipality, the different types of soil encountered are as follows:

Deck Dior soils, more common and more represented in the city, are tropical loam soils with a high silt and clay content. They also have a high water retention capacity. These soil types occupy the central and peripheral parts of the city. With a high clay content, the soil quickly becomes saturated, causing water to stagnate and accentuating flooding throughout the town during the winter months. These are soils suitable for farming and capable of providing excellent crop yields in a year of normal rainfall. In addition to this soil category are calcimorphic soils, which are Deck Dior soils with a very high limestone content. Their texture is more compact than Deck Dior soils. According to the Reference Town Planning Scheme, they are found almost everywhere in the Wakhaldiam, Leona North and Leona South districts.

The Dior soils found in the north of the town and on the outskirts of the Wakhaldiam and Leona Nord districts are tropical ferruginous soils with low mineral and organic content. They are very leached and degrade very quickly due to the effects of monoculture in this area and climatic conditions. These soils are ideal for urban development and construction (PUR).

Rainfall is unevenly distributed across the globe. Some regions receive much greater amounts at a time when some are deprived of rainfall and find themselves in a situation of extreme drought. The amounts of rain seen in the department of Bambey and collected from one year to the next vary from 400 to 800 mm per year. Under the influence of the predominantly Sahelian Sudano-Sahelian climate, Bambey has a low rainfall compared with the regions to the south of the country.

Observations made at the Bambey station between 2007 and 2010 show a rainy year (2008) with more than 800 mm of water (ANSD/SRSD/DRDR).

In the department of Bambey, as in the town of the same name, temperatures are generally high and vary from an average low of 24°C in January to an average high of 38°C in May-June. On rare occasions, however, the minimum temperatures drop, accentuating the daily temperature range. These moments are generally observed during the first quarter of the year with mild temperatures, but with the arrival of the harmattan (February-March) this trend is reversed very quickly before the arrival of the monsoon between July and September (wintering period).

1.3 A predominantly agricultural economy

A country town, rural town or farming town, Bambey is one of those towns in the country where the farming population is the largest. This justifies the predominance of agriculture and livestock farming in the town. However, the development of commerce remains a reality in the town, even if it is dominated by small-scale, informal trade. In addition to these sectors, there is also the craft industry, which is not very dynamic but which, despite everything, is becoming the third largest sector of activity in the town.

In the following lines, we have analysed the different sectors of economic activity in the town to get a clear picture of its economic development.

1.3.1 Basic sectors of activity (agriculture and livestock)

These two activities constitute the main sectors of activity in the department in general and occupy an important place in the town's economic activities. In fact, the town's primary vocation and its

geographical location, at the heart of the groundnut basin, mean that agriculture is a key activity that contributes to food production and household income.
The town owes its development to the expansion of groundnut cultivation and the existence of the railway in colonial times, as it was a place where groundnuts were marketed, making Bambey a stopover town. Today, despite the decline in arable land due to urbanisation, the degradation of agricultural land and the drop in rainfall in recent years, the town still owes its economic influence to agriculture and livestock farming. The livestock market, which takes place every Friday and attracts traders from all over Senegal, is the biggest and most popular market in the department, even though most of the traders come from Diourbel, Linguere, Khombole and the rural communes of Bambey, such as Ngoye, Ndangalma, Thiakhar, Refane and Ndolndol.

1.3.2 Trade, a sector dominated by the informal sector

Little developed because of the town's poverty, trade is one of the most dominant economic activities in the town after agriculture and livestock. Bambey's central market is the place of choice for most commercial activities. Informal trade also accounts for a very large part of the sector. It is mainly located in and around the bus station (PDU de Bambey horizon 2021). These floating businesses employ a large proportion of housewives, a source of secondary economic activity.
Trade is experiencing a new dynamic thanks to the development of weekly markets in the department. This is a fact that is particularly noticeable in the rural communes, and confirms the once-polarising role of the town of Bambey. The small number of large traders, the proliferation of small businesses and, above all, the holding of a specific livestock market all point to the ruralization of the town's functions.
For the former rural communities, the establishment of weekly markets represents a major challenge in the quest for increased revenue. The department has thirteen weekly rural markets (PDU). The dynamism of trading activities is also boosted by neighbouring weekly markets such as Mbafaye, in the department of Mbour, Niakhar, in the department of Fatick, Touba-Toul in the department of Thies, and even the national market at Dahra, in the department of Linguere. The scale of these markets can be seen even beyond the national territory, as many Malians, Gambians and Mauritanians sell or buy products from these markets (PDU).
The popularity of weekly markets is real, and they are helping to improve local authority revenues and diversify people's activities.

1.3.3 The craft industry, a sector in decline

The craft industry is the sector of activity in the town that brings together a good number of working people, despite the problems in this sector. In fact, most of the people working in this field are in the informal sector. From woodworkers to dressmakers, the sector can be divided into two. Firstly, there is the production craft industry, which includes woodwork and joinery (wood and metal). Secondly, there is the service sector, which includes mechanics. But the difficulties encountered in this sector include a lack of vocational training and a lack of awareness of the sector in the town. Arts and crafts are not widely practised in the town, apart from by housewives in the neighbourhoods.
Despite the large number of people working in this sector, the craft industry is still an unregulated and little-known sector of economic activity in the town. As a result, most people working in the sector consider it to be a secondary activity, as do housewives.
The town of Bambey lies between latitude 14°42 North and longitude 16°27 West. It shares its communal boundaries with four other communes (Ngoye, Ngogom,
Thiakhar and Ndangalma). The physical environment has three types of soil. These are dior soils, deck dior soils and callimorphic soils. The physical environment is conducive to flooding, as the soils are almost entirely impermeable.
Bambey's economy is dominated by agriculture, despite the development of other economic sectors such as trade, even though the latter is characterised by informality. The craft industry employs few people in the town. But beyond these sectors of activity, we can cite the development of the digital sector with money transfer services that are attracting more and more people to the town.

But whatever the economic situation of the town and the characteristics of its physical environment, Bambey is today a dynamic town that has been attracting people ever since it was founded.

Chapter II: Urban dynamics in the town of Bambey

The town of Bambey was granted full municipal status in 1956, and continues to attract people. With its many infrastructures and facilities, it is the focal point of an entire department. Before independence, the marketing of Karachi and the railway station played a major role in the town's development. But in the aftermath of independence, the city saw new economic activities develop, attracting more people and continuing its expansion into the surrounding villages. And this despite the natural, legal and administrative constraints of the area.

In this chapter, after a brief summary of its spatial organisation, we attempt to describe the spatial dynamics of the city in the post-independence period, from 1960 to the present day. In this section, we focus on three periods. The first is the period from 1960 to 2000, followed by the period from 2000 to 2010, marked by the first political changeover in Senegal, during which the town of Bambey benefited from a number of favours due to the fact that the mayor at the time belonged to this political movement, and finally the period from 2010 to the present day, marked by increasingly intense urban growth.

11.1 Spatial organisation of the town of Bambey

The town of Bambey covers an area of 7 km^2 . It is divided into two parts, north and south, by the railway leading to Touba. In colonial times, the town was divided into two districts: the settlers' district (Escale), to the north of the railway, and the indigenous people's district (DVF) to the south. This was due to the important role played by the railway station before independence. Today, the town is divided into four districts: Wakhaldiam, Leona North, Leona South and DVF.

According to the town planning master plan for the town of Bambey, the department's main roads running through the town were used by the authorities to divide the municipality into four districts. To the north of the main road are the Wakhaldiam neighbourhoods (120ha),

Leona North (181ha) and Leona South (119ha), while in the south there is a single district, D.V.F, covering around 500ha.

Map 2: Spatial organisation of the town of Bambey

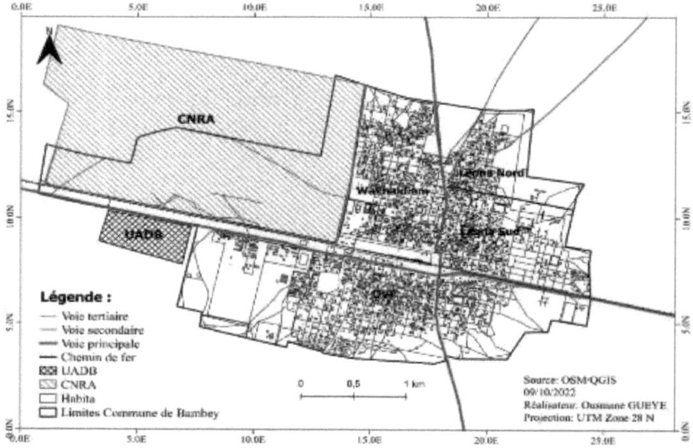

Source: OSM/Google Earth/QGIS

11.2 Urbanisation process in the town of Bambey

Starting in 1960, the town's urbanisation process can be divided into three phases. These different phases are described in the following lines in order to understand the genesis and the process by which

Bambey has become a university town in the middle of the countryside.

II.2.1 The period from 1960 to 2000: urbanisation around major transport routes

From 1960 until the 2000s, urban development in the town of Bambey was very limited. And if it existed at all, it was confined to the main roads. This is known as linear urbanisation. In fact, population concentrations were limited to the areas around the DVF, the indigenous people's quarter, and Wakhaldiam, the settlers' quarter, the urban centre. These districts are separated by the RN3 and the railway, creating a socio-spatial segregation in the occupation of space.

The town of Bambey originally developed on either side of the railway, with the settlers' quarter to the north, known as "Escale Commerciale", and the indigenous village to the south of the railway, which is now the DVF quarter. For a very long time, therefore, the railway has been the structuring axis that has influenced the spatial organisation of the town, as it has supported the flow of people and goods arriving in or passing through the town of Bambey, whose only interchange centre was the railway station (PDU).

As the capital of the department, the town is starting to look really attractive, thanks to a number of administrative services. In fact, the urban dynamic began to take shape mainly at this time in the history of Senegal, and of the commune in particular. These services, such as the police, the gendarmerie, the prefecture and the departmental court of Bambey, as well as the health and education infrastructures, are points of attraction that enable the town to polarise an entire department. This is not an asset of the town of Bambey as such. However, it is a question here of a political will manifested by the political authority with a vision for the organisation of the national territory. The town's development was strongly influenced by the groundnut trade at the time. It was like the trading towns, which "*constituted trading posts where one or more traders were based*". These types of town "were *generally organised around the area adjacent to the railway station. By a knock-on effect, related activities linked to the marketing and evacuation of groundnut products were set up" (*DIOP A., 2005).

The town's territorial organisation, inherited from the colonists, meant that everything was organised around the "Escale" district, and continues to be so to this day. In fact, the escale is the vital centre with its various facilities. The space is organised concentrically, and the urban fabric becomes denser the further one moves away from the urban centre, the centre of business, administrative buildings, roads and various networks (DIOP A., 2005). This segregative logic of the town established by the colonists can be explained by the fact that it developed along the railway and the RN3. These two communication routes divide the town into two parts (North and South).

As in all countries, the drought of the 1970s contributed to the spatial development and demographic weight of the town. Indeed, marked by agricultural crises (FAYE S., 2018), the drought accentuated the displacement of rural populations from the surrounding countryside to Bambey. The town, a beacon of hope for a population hit by the hunger gap, the agricultural crisis and famine, has seen its population grow overnight as a result of the agricultural crisis in the surrounding countryside. Rural land became infertile, agricultural harvests fell dramatically and desperate people left the countryside for the cities. This led to the need for new land for housing, which in turn led to the expansion of the town or the spreading of its urban fabric onto farmland: this was rurbanisation.

This movement of people from the countryside to Bambey, combined with the town's natural demographic growth, meant that its population rose from 5,600 in 1966 to 9,853 in 1976 (FAYE S., 2018). According to the Direction des Statistiques et de la Prevision (DSP). An area of 152 hectares was urbanised between 1960 and 1997[17] . This can be explained by the increase in the city's population. This urban expansion of the city is justified by an urbanisation rate that was respectively between 6.3% in 1979 and 8.42% in 1988 (ANSD, 2015)[18] .

Map 3: Land use map of the town of Bambey in 1990

---

[17] Cited by FAYE M., 2013. L'implantation de l'universite Alioune Diop de Bambey comme facteur d'urbanisation, Master's thesis in geography.

[18] Economic, Social and Regional Situation of the Diourbel Region in 2015

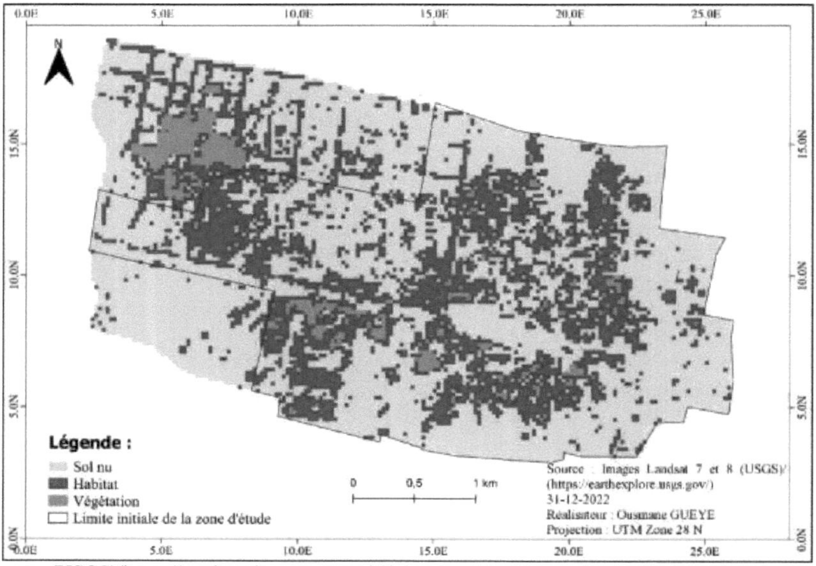

Source: (USGS)/https://earthexplore.usgs.gov/)/QGIS

This 1990 land use map shows that despite Bambey's relatively high population density, empty land occupies 8.45km^2 of the city's surface area. In principle, significant population growth consumes space. It will therefore be interesting in the following lines to see what the surface area of the built-up area and bare land in the town would have been over the next 15 years (2005), given the high demographic weight of the town.

11.2.2 The era of controlled urbanisation (2000 to 2010)

In the 2000s, the town of Bambey experienced a new spatial dynamic, with its population continuing to grow. In 2002, according to the RGPH, the town had a population of 20,366, compared with 15,766 in 2001. The town's urbanisation rate rose from 8.42% in 1988 to 9% in 2002. This meteoric population growth is not without consequences, as it has an impact on the need for housing space. As a result, several housing development projects have been initiated by the local authorities to meet the growing demand for housing.

Around 2007, the town of Bambey gradually outgrew its original communal boundaries and spread into the national domain land belonging to the surrounding rural communes. In this way, we can see the development of the built-up area to the north towards the rural community of Ndangalma, to the east towards Thiakhar and, above all, to the south towards Ngoye. With the establishment of the regional university college in the commune, which was created in 2004 and became a full-fledged university in 2010, and which is located within the communal boundaries of Ndangalma, we are witnessing the development of new buildings and the advance of construction in this area, which extends overnight towards the village of Bambey Serere. It was also during this period that the urban sprawl of the town began to be increasingly noticeable, confrontational and worrying. The establishment of Bambey's Alioune Diop University and the new activities that were emerging in its wake accentuated the need for housing, with the result that the race for property became a reality that had previously gone unnoticed in the commune. The development of the property sector is becoming an alternative way of meeting the many housing demands.

The expansion of the town was reflected in the fragmentation of the agricultural fields of the farmers of Keur Sakoura Badiane, a hamlet in the Leona Sud district of Bambey. At the time, all the plots were negotiated on land belonging to the national estate, which was occupied and developed by farming families.

The following map shows the city's land use in 2005.

Map 4: Land use map of the town of Bambey in 2005

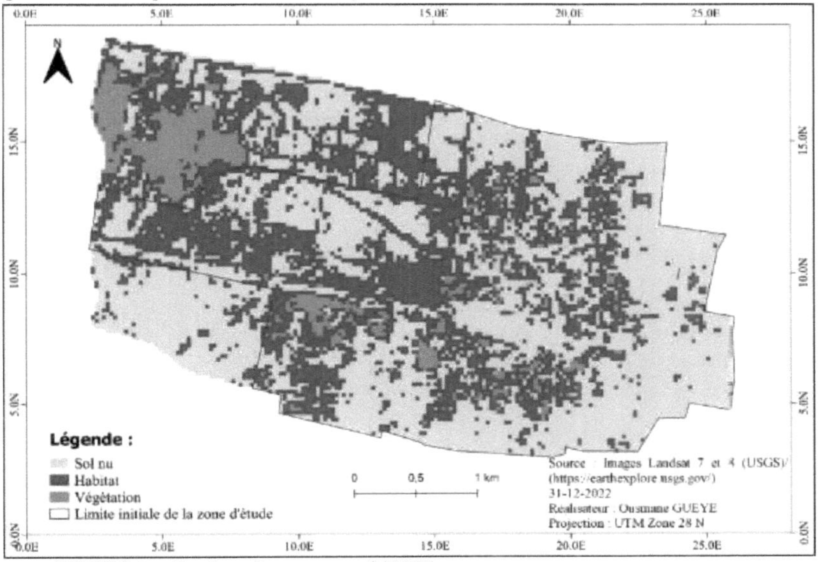

Source: (USGS)/https://earthexplore.usgs.gov/)/QGIS

This map shows a reduction in the area of empty spaces in favour of buildings and vegetation. The built-up area, which stood at 4.35 km^2 in 1990, was 5.11 km2 in 2005, while the area of empty spaces or land decreased from 8.45 km2 in 1990 to 7.32 km2 in 2005.

This analysis shows that as the population increases, the built-up area expands, the city spreads out and the empty space decreases in size. So, in the next 15 years (2020), what will the town's situation be in terms of the occupation of its municipal perimeter?

11.2.3 From 2010 to the present day, an era of intense and uncontrolled urbanisation

In the aftermath of the 2010s, the town's population grew to 28,908 (RGPHAE, 2013). The period from 2010 to 2020 is truly the era of urbanisation for the town of Bambey, coinciding as it does with the full operation of the Alioune Diop University. The creation of new infrastructure and facilities in the town is attracting more people from the countryside, from the former rural communities of Ngoye and Thiakhar, among others. As a result of this strong natural demographic growth and population flows, the city's urbanisation rate rose from around 9% in 2002 to 10% in 2013 (RGPHAE, 2013).

This strong demographic growth is unevenly distributed across the city's different neighbourhoods. In fact, there is a high concentration of people in the DVF district, moving towards Ngoye in the south, towards the Bambey-Fatick road, and the development of the built-up area with new constructions confirming the urban expansion of the town towards agricultural land. These new buildings are sometimes the result of a family from the surrounding countryside who, fearing that their field would be reclaimed and subdivided by the municipality, decided to take steps to build their land and make it available to tenants. The head of Bambey's town planning and housing department says: "*It's the villagers who are investing in property and land. In fact, they are the owners of the agricultural fields on the outskirts of the town, if we refer to customary law*" (interview, October 2022).

The development of buildings on agricultural land is juxtaposed with the increase in demand for housing. In fact, the arrival of students and civil servants in the town every year creates a need for housing, as does the natural demographic growth of the town's population. These facts, both natural and man-made, are contributing to the boom in the property sector and in rental values (interview with the head of the town hall's technical department, August 2022).

Decree no. 2003-410 of 04 June 2003, which approved and made executory the master urban development plan for the commune of Bambey, means that today, despite the legal and administrative constraints on its expansion, the town of Bambey already has land reserves that enable it to meet the population's demand for housing, as well as basic facilities and social services. Today, the town is expanding almost entirely to the south and east, with housing development projects planned for 2020 in the DVF and Leona neighbourhoods. These developments, which were simply extensions of these neighbourhoods, have fragmented what was left of the town's land reserves. However, it should be noted that the development of buildings and informal settlements on the outskirts of the town is aggravating the situation, as it is causing land insecurity with multiple consequences. Indeed, following decree n°2011-1991 of 16 December 2011 amending and extending the territorial boundaries of the commune of Bambey, the authorities of the town of Bambey, in 2020, wishing to initiate subdivisions on land allocated to their commune by the State, came up against fierce opposition from the municipal authorities of the four communes surrounding the town.

Figure 1: Trends in the city's urbanisation rate from 1979 to 2013

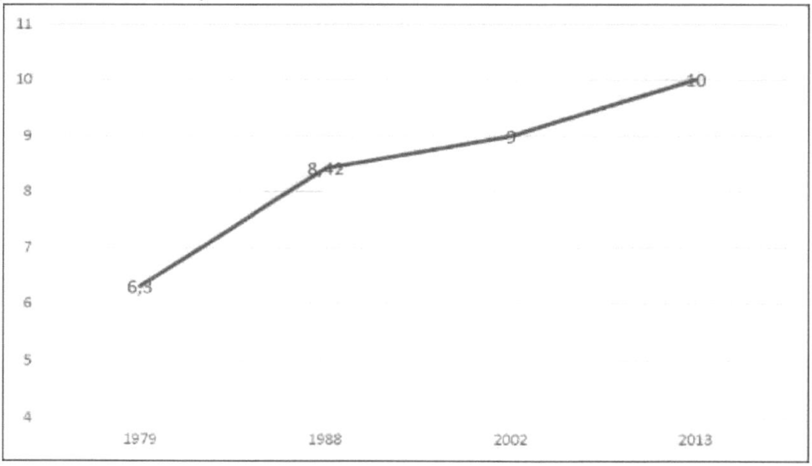

Source: ANSD

This graph shows the evolution of the city's urbanisation rate from 1979 to 2013. It stood at 6.3% in 1979 before reaching 10% after more than 34 years in 2013, rising to 8.42% in 1988 and 9% in 2002. From this graph, we can see that the rate of urbanisation in the city is increasing by more than 1% every 10 years.

Map 5: Land use map of the town of Bambey in 2020

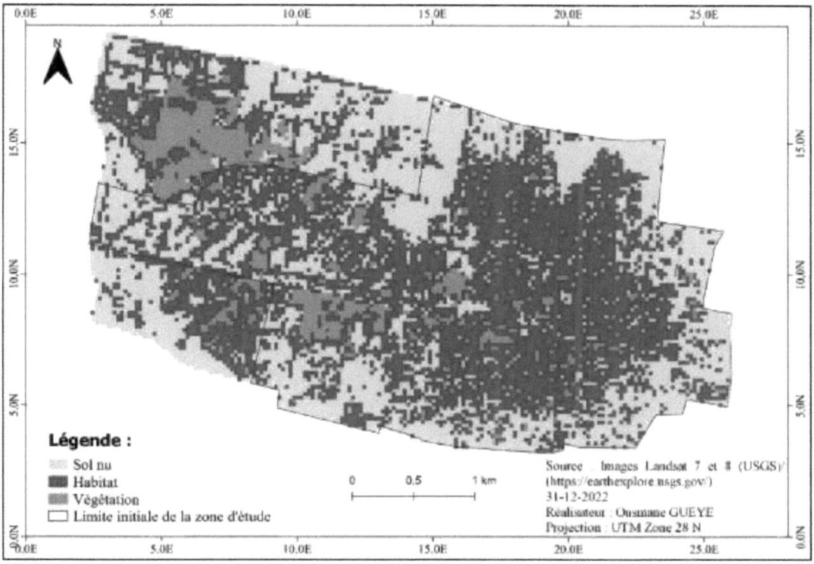

Map 6: Summary map of the spatio-temporal dynamics of the town of Bambey

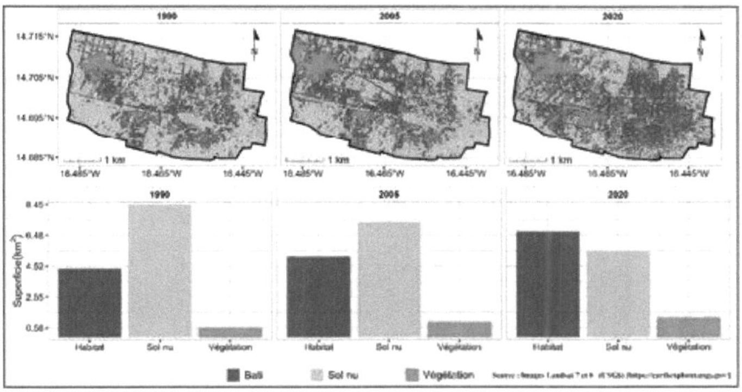

Source : USGS/httpas://earthexplore.usgs.gov/

The spatial evolution of the town of Bambey has become a remarkable fact. Over the last 30 years, the built-up area has grown steadily and the agricultural land around the town has become increasingly sought-after. The following graph shows how the surface area of the town has changed over the last thirty (30) years.

Figure 2: Change in surface area of buildings, vegetation and bare soil from 1990 to 2020

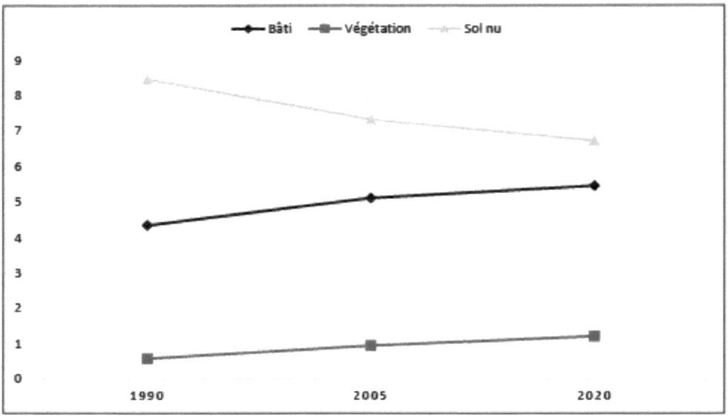

This graph shows that the surface area of vegetation and buildings continues to increase every fifteen (15) years, unlike bare land, whose surface area is constantly decreasing to the benefit of buildings. In fact, the built-up area measured in 1990, 2005 and 2020 was 4.35, 5.11 and 5.45 km respectively[2] . The area of vegetation is 0.58, 0.95 and 1.21 km. On the other hand, the surface area of empty spaces or bare ground was 8.45 km2 in 1990, 7.32 km2 in 2005 and 6.72 km2 in 2020.
This spatial development of the city since its birth is the result of a combination of several factors in time and space. In the next chapter, we look at the various factors that accentuate the city's spatial dynamics.

Chapter III: Spatial dynamics of the town of Bambey

This chapter looks at the different factors in the spatial dynamics of the town of Bambey, based on the results of our research (documentation and field survey).
We have therefore endeavoured to identify the main factors behind urban growth in order to gain a better understanding of the spatial revolution in the town of Bambey.

111.1 Dazzling demographic growth since 1976

The town of Bambey, like many of the country's second-tier towns and the capital of a department, is undergoing major social and spatial change. These social and spatial changes are characterised by significant demographic growth and spectacular urban dynamics.
The evolution of a town's population is a key factor in driving its spatial dynamics. In fact, population growth always leads to a need for housing, which in turn consumes space and therefore drives spatial expansion. In fact, this population, which increases year on year as a result of the natural demographic growth of the department and the effects of migration, puts a great deal of pressure on land, whereas most of the remaining land reserve is made up of farmland and grazing land. Local public authorities therefore need to find space to house the population and provide adequate infrastructure and facilities to support the town's spatial and economic development.
Following the administrative reorganisation that divided the Louga and Diourbel regions, Bambey became the third department of the Diourbel region, alongside Mbacke and Diourbel. A department with significant demographic growth, marked by an unequal distribution of the population between rural and urban areas. In fact, the whole of the department of Bambey is made up of a rural population, with 90.3% of its population living in rural areas (ANSD, 2013).
According to the 2013 RGPHAE, the department of Bambey is home to 20% of the population of the Diourbel region, behind Mbacke with 62% and ahead of Diourbel, which comes third and last with 18%. However, despite this dichotomy between rural and urban dwellers in the department, the town of Bambey is experiencing impressive urban growth, reflecting the dynamic of its urbanisation since the post-independence period.

According to DPS data, the town's population almost doubled between 1960 and 1997, rising from 5,600 in 1960 to 9,853 in 1976[19]. The town of Bambey is experiencing strong demographic growth, which is taking up a lot of space, with a relative decline. The average rate of urban population growth was 3.4% in 1961 and 1976. However, it has to be said that this rate is lower than the average observed in the country's communes at the time, which was 4.9%.

Between 1976 and 1988, the city's average annual growth rate was 4.3%. This was higher than the growth rate in the city during the 1960s. However, this population growth was to start to decline again in 2002. This is justified by its growth rate, which was 2.5% (RGPH, 2002), followed by an increase of 3.7% between 2005 and 2015. In 2016, the urbanisation rate for the town of Bambey was 9.7%, compared with 16% for the Diourbel region (ANSD, 2016).

This high demographic growth led in turn to rapid and continuous urban expansion. In 2002, the population of the town of Bambey was 20,366 (RGPH, 2002). This demographic weight, characterised by a young population with 55% under 20 years of age, creates a demand for housing that will have to be met in the coming years or decades, despite the exhaustion of the town's communal boundaries.

This upward demographic growth reached 28,908 inhabitants according to the latest population census in 2013. This growing population is a major factor in the city's current urban dynamic. The ANSD's population projections for Senegal ([19 20]) show that the urban population of Bambey will reach 39,474 in 2023, 40,753 in 2024 and 42,077 in 2025.

This demographic growth of the city cannot be considered by taking into account only the natural growth of the population. In fact, the phenomena of temporary or seasonal migration by students and other civil servants in the city cannot be left unaddressed or reduced to nothing, as they constitute an extremely important part of the city's population and have huge consequences in terms of consumption of space, which translates into a demand for housing, given the inability of the social campus to accommodate them.

Figure 3: Population growth from 1988 to 2025

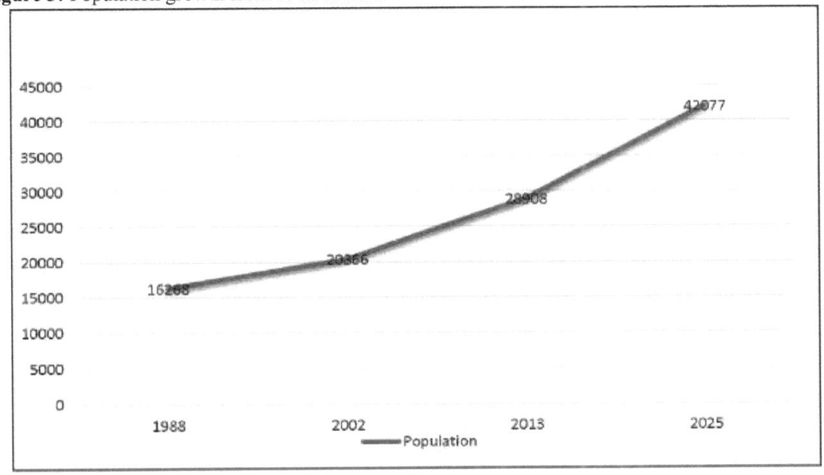

Source : ANSD

This ever-growing population is distributed unevenly across the city's four neighbourhoods. The DVF district remains the most populous, with 11,146 inhabitants, or 38.53% of the town's total population.

[19] Quoted in FAYE, 2013. L'implantation de l'universite Alioune Diop de Bambey comme facteur d'urbanisation, Master 2 thesis in geography.

[20] ANSD, 2013: Population projection for the Dakar region

It is the working-class district that is home to the majority of the city's farming population. The Leona North district is the least populated, with 5,113 inhabitants, or 17.7% of the total population. Leona Sud has 6,813 inhabitants, or 23.57% of the population. The Wakhaldiam district, on the other hand, is home to
5,836 inhabitants, or 20.2% (ANSD, 2013). The distribution of the town's population shows disparities. In fact, dividedẽe into four major districts, the town of Bambey concentrates the largest part of its population in the South in the DVF district.
The following graph shows the breakdown of the city's population by district.

Figure 4: Population band diagram by neighbourhood

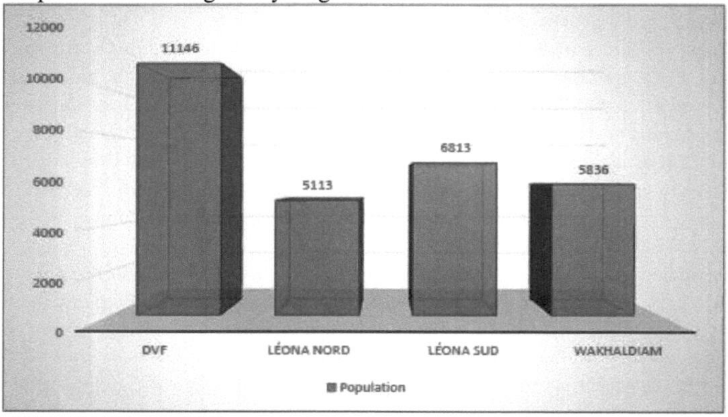

Source: ANSD, RGPHAE, 2013

Table 5: Breakdown of the city's population by gender and district

Neighbourhoods	Men	Women	TOTAL
DVF	5 296	5 851	11 146
Leona Nord	2 331	2 782	5 113
Leona South	3 182	3 631	6 813
Wakhaldiam	2 770	3 067	5 836
TOTAL	13 579	15 331	28 908

Source: ANSD/RGP] HAE, 2013

With an ever-increasing population and estimates that show no downward trend in the coming years, the town will continue to be subject to this ever-increasing demographic pressure. This suggests that, despite the administrative and legal constraints, the initial municipal boundaries will be extended in the future. This demographic boom is creating and will continue to create enormous pressure on the agricultural land of the surrounding municipalities, and will therefore cause social, economic and environmental upheaval in the town and its surroundings.

Migration is a remarkable phenomenon in the commune. However, in contrast to what is happening in some of the country's cities, with international, national and even regional migration, Bambey is experiencing a population shift towards the country's major cities, such as Dakar. However, it is also important to note the large number of people who come to the University and other vocational training establishments year after year. These migrants make up a significant proportion of the population, which increases the demand for housing in the city.

According to FAYE M. (2013), the town's demographic growth is due as much to the increase in migrants to the town as to the fertility of the population. In fact, most of the migratory population in the city is attributed to the rural population wishing to take advantage of the benefits of the urban environment. These advantages can be summed up as quality education, access to quality healthcare and leisure services.

Although demographic growth is one of the key factors in the city's urban growth, as it is at the root of rapid and uncontrolled physical urbanisation, the city's facilities, such as the university, the central market and the health centre, which give the city command over its territory, are also important factors in the city's development. Depending on its attractiveness, the town has a major influence on the rest of the department. As a result, the town's attractiveness and ever-expanding area of influence are key factors in its urban dynamics.

111.2 Bambey, a town that controls its territory

Bambey is a dynamic town in terms of research, training and the economy, and its influence extends throughout the country. In fact, the presence of training and research infrastructures such as the National Agricultural Research Centre and facilities such as the market, as well as its health district, give the town a level of influence and polarity over all the communes in the department. But this influence is not confined to the department itself, as it extends beyond departmental, regional and national boundaries. In fact, thanks to the National Agricultural Research Centre and the university, the town polarises all the other regions of Senegal, attracting more and more students and researchers. Despite its low population, the town's appeal remains significant.

On a departmental scale, the town of Bambey makes its appeal and its power felt over a radius of more than 35 kilometres, as it has political, administrative and economic command over all the other communes in the rest of the department. The relationship between the town and the rural communes of the department means, firstly, that the urban area is dynamic and driven by the flow of people and goods in the context of trade, as it is the place where agricultural produce from the countryside is sold and, by a knock-on effect, offers the countryside a number of consumer goods that can only be found in the town. Indeed, "*the level and intensity of urban-rural relations is the most important barometer for assessing the city's functionality in relation to its environment*" (DIOP A., 2005). Thus, the driving forces and mechanisms behind the attractiveness of cities derive from the powers of command and the quality of the services offered to the environment (DIOP A., 2005).

The town's infrastructure and facilities play a vital role in its spatial and economic development. In fact, the town's university and health district are the reference points for the daily influx of people. These structuring poles, capable of attracting a huge population, remain the main infrastructures of the town and even of the department. They have a great capacity to influence the rural communes which, along with the commune of Bambey, make up the department. Their polarity can to some extent be seen at regional, national, sub-regional, even continental or global level, in the case of the university and the CNRA.

Between 2000 and 2010, the Lycee de Bambey was to some extent the only secondary school in the department, and pupils came from all the rural communities in the department. However, since 2012, with the creation of local high schools, we have seen communes that had no high school before that year end up with more secondary schools than the town of Bambey. One example is the commune of Ngoye, which has two secondary schools, as well as the new Bambey Serere secondary school. However, communes such as Thiakhar find themselves without a secondary school, and their pupils are forced to commute between Thiakhar and other secondary schools in the department. However, it should also be noted that the Lycee de Ngoye is now the school of choice for these pupils.

111.2.1 Bambey, a town with a polarising medical plateau

The health map of the department of Bambey shows many spatial disparities in its distribution and organisation. As the capital of the department of the same name, the town of Bambey benefits from medical and social facilities, although the standard is not met, as it is well below the threshold set by the WHO. With a population of 29,908 (ANSD, 2013), the town of Bambey has a health centre, a health post and a dispensary in terms of public health facilities, the main structures for receiving patients (PDU de Bambey).

However, it should be noted that despite their inadequacy, these health establishments are the pride of an entire department, because the town, with its health centre, has the highest level of health equipment in the whole department. All the other communes, even those with more than 40,000 inhabitants, such as Ngoye[21] , have a health post and some have a health centre. As a result, most of the department's patients are referred to the Bambey health centre by the head nurses.

As a result, the medical facilities in the town of Bambey are unable to provide care for the entire population of the department, as the demand for healthcare is very high. This means that the health facilities in Diourbel and Touba have to take up the slack, as they are the reference health establishments from an administrative point of view, and are therefore the preferred relay or reception facilities.

The Bambey health district covers 25 rural health posts and 3 urban health posts (Leona Nord, Croix Rouge and Saint-Martin). It serves the entire population of the department, which totals 299,477 inhabitants (RGPHAE, 2013) and covers an area of 1,335 km^2 . This vast area is not without consequences for the population's access to healthcare, as some localities have to travel distances of up to 35 km. Added to this is the fact that most of the roads linking Bambey and the other communes are sandy, and the main means of transport is still horse-drawn, apart from the ambulances that transport the sick. This geographical accessibility poses real problems and has the same impact on people's health as economic accessibility. This is one of the reasons why the department is so poor, and why the means of transport are so limited.

In terms of the spatial distribution of the population in the department, the commune of Ngoye accounts for 15.42% of the population of the health district, followed by the commune of Refane with 11.26% and the commune of Ndangalma with 10.90%. However, this unequal distribution of the population in the different communes cannot be attributed to the distribution of health equipment at departmental level. While WHO standards require one health post for every 10,000 inhabitants and one nurse for every 300 inhabitants and one midwife for every 300 women of childbearing age in rural areas, the commune of Ngoye, with a population of 46,174, has just one health post and two health huts, in the villages of Battal and Gourgourene respectively. The situation is similar in the commune of Ngogom. These disparities in population distribution

by health post show that there are very important services available and make us think about the quality of the health care received by the populations in need.

The commune of Bambey, with a smaller population compared to the rural communes of Ngoye, Refane, Ngogom and Ndangalma, i.e. 9.98% of the district's total population, has more health facilities. The population is divided between the health centre and the town's three urban health posts. As a result, even though it has a smaller population than the other communes, it stands out from the crowd with more facilities and structures to care for patients. This is understandable, as it is the capital of the department, and therefore the only town in its entire territory.

The remaining communes, with low representation rates of less than 10%, are characterised by a low presence of health facilities and a low population compared with the communes of Ngoye, Ndangalma, Refane and Bambey. However, it should be noted that the physical accessibility of health facilities due to the layout of dwellings (scattered settlements in rural areas) poses real difficulties in communes such as Ngoye, where people have to travel more than 5 kilometres to seek care. As a result, the supply of and access to healthcare are becoming problematic, and the physical and economic accessibility of health facilities are the real problems in terms of health.

Figure 5: Breakdown of the health district's population by commune

[21] Ministere de l'economie et des finances/Agence nationale de la statistique et de la demographie, 2005. *Situation economique et sociale du Senegal.* WHO standards for health infrastructure are one (1) hospital per 150,000 inhabitants, one (1) health centre per 50,000 inhabitants and one (1) health post per 10,000 inhabitants.

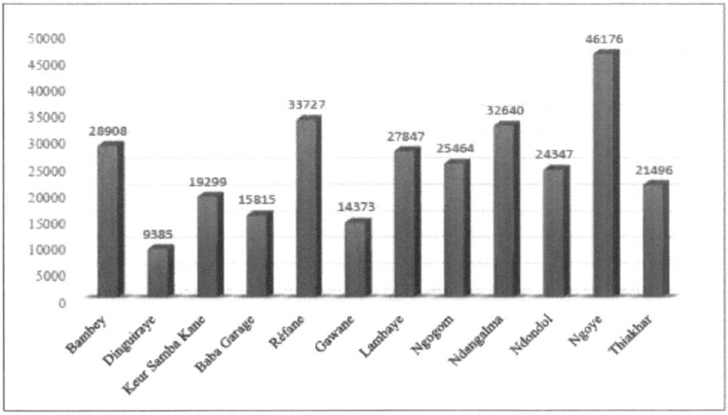

Source: ANSD, RGPHAE, 2013

The area polarised by the Bambey health centre has a radius of up to 35 km. This shows that despite the distance, people continue to visit the health centre. However, there is no shortage of people who, from an administrative point of view, belong to the Bambey health district, and the health centre is their first point of contact. But with distance (physical or geographical accessibility), the health structures of Khombole, Tivaouane and Thies become their places of frequentation. In terms of the quality of healthcare on offer, the Touba and Diourbel health districts are also increasingly attracting the population of the Bambey district.

The following map shows the area polarised by the Bambey health centre, which extends to a radius of 35 km. The populations of the communes of Keur Samba Kane and Dinguiraye are more affected by the distance-cost ratios from a physical and geographical point of view.

Map 7: Polarised area of the Bambey health district

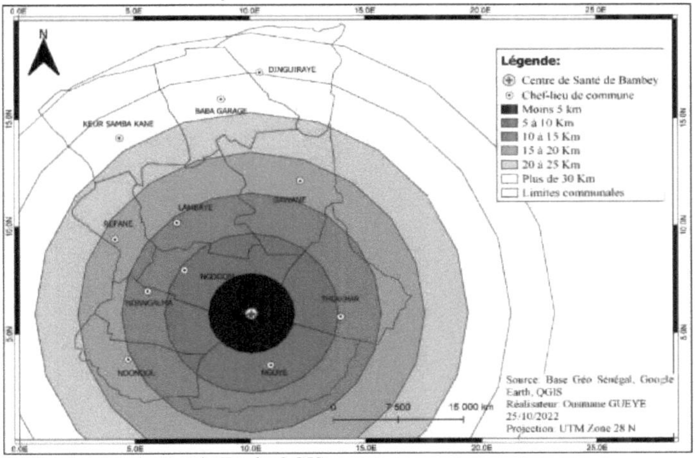

Source: Base Geo Senegal, Google Earth, QGIS

The Bambey town health centre is the highest level health facility that the people of the 12 communes in the department can expect to find quality health care in the event of an extreme emergency or serious illness.

With a population of 299,477 (RGPHAE-2013), the department of Bambey is faced with a real headache when it comes to providing and using healthcare. Health centres such as Ngoye, with a population of 46,176, are unable to provide care for all these people using a single health facility, staffed by a single resource person, even though some communes have set up health huts to provide this service. But it has to be said that most of the health centres are in a very advanced state of disrepair, while others have not yet been completed.

As a result, the health centre, the health post and the Red Cross dispensary in the town of Bambey are facing a huge influx of people from the department, especially from rural areas.

Like the Bambey health centre, the town's market also contributes to the town's attractiveness by accentuating town-country relations, especially in the post-winter period.

111.2.2 The central market in Bambey

The development of commercial activities is a fact that has always accompanied the urbanisation of regions. The world's largest cities are now home to major markets. In secondary towns, such as Bambey, the market concentrates a variety of products and plays a very important role in influencing the local area, accentuating relations between the town and the rest of the department. The dynamics of the flow of people and goods between the town of Bambey and the other communes in the department are sufficient evidence of the town's attractiveness and its influence on the rest of the department. The dynamics of the town's central market today reveal the importance of trade, which also fuels the town-countryside relationship between the town of Bambey and the former rural communities. According to FAYE S. (2018), the surrounding countryside is Bambey's granary.

Like the town's other facilities, the Bambey market is one of the key elements in Bambey's urban development, even if the environmental consequences are now a real problem.

Today, more than 50% of the products sold in Bambey's market come from outside the department, and most of the traders are not from Bambey. The market therefore plays a vital role in the town's economic dynamic, and we can conclude that it is the essential element in the town's economic dynamic.

Thanks to its market, the town's area of influence extends far beyond the department. It extends to more than 50 km from the town of Bambey.

Map 8: Polar area of the central market in Bambey

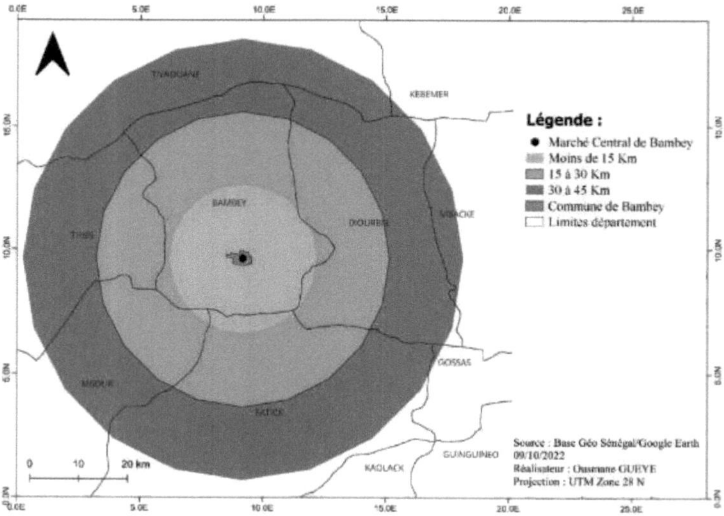

Source: Base Geo Senegal/ Google Earth

111.3 The University of Bambey, a real factor in the current urban dynamic

The creation of the CUR in 2007 in the town of Bambey, in turn, encouraged the development of buildings to the west of the town, towards the village of Bambey Serere. The university is located on national estate land belonging to the former rural community of Ndangalma, now the commune of Ndangalma. The creation of the university in 2010 and the arrival of baccalaureate holders year after year from all over the country have given the town of Bambey a national dimension.

The development of buildings, especially modern constructions in this part of the city, has led to speculation over land in this area. Farmers, who own the land under customary law and the system of inheritance, have seen their arable land shrink in favour of the built-up area. As a result, they tend to divide up their fields and sell them to private owners for housing, out of fear that the State or the local authority will take them over for public use.

This socio-economic conversion in this part of the city, whose economy was once based on agriculture and livestock farming, has become a reality with the emergence of new economic activities such as the development of small businesses around the University. There is now a huge demand for housing throughout the town, which explains the growth in the built-up area. This has enabled the city to fulfil its function of housing its population, in accordance with the Athenian charter, which states that a city must fulfil four functions: eating, sleeping, moving, housing and working.

The university's impact on the town's urban development has been remarkable. According to the head of the technical department at Bambey town hall: "*Thanks to the creation of the university in 2010, Bambey has become a city of education and research. With the establishment of the university, we've also seen a property boom, the construction of modern buildings, and the occupation of a large area of land, farmland belonging to farmers in the surrounding villages that had long been left empty during the dry season*".

The rush of private individuals into this part of the town has had an impact on the price of land, but also on all the land in the town. Today, we see people leaving the communes of Ngoye, Thiakhar and Ndangalma, formerly owners of fields, coming to sell land that, according to customary law, belongs to them because they have always farmed and exploited it. But with the decree extending the territorial limits of the town of Bambey, these lands are now part of the land base of the town of Bambey. This has given rise to disputes that are sometimes complicated, even leading to demonstrations by local people, sometimes against subdivision projects planned by the town of Bambey. The soaring cost of rent and the high price per square metre mean that land disputes are very common in this area, sometimes pitting Bambey town council against farmers from the surrounding communes. In fact, "*land speculation in this part of the town is giving rise to radical feelings on the part of the farmers, the former owners of this land. The rise in rental costs and rental values, and the skyrocketing price of m^2 for plots of land mean that the whole of this part of the town has been broken up into plots for residential use, with 70% of the plots already occupied and the rest under construction. As a result, the town has devoured its entire land base, and the phenomenon of rurbanisation has become a real fact of life in this part of the town, as we are witnessing a double cohabitation in this area with the new inhabitants of this part and the inhabitants of the village of Bambey serere*".

The ruin of the agricultural sector in the area and the development of the real estate sector with its occupation of space mean that the production of used water and waste poses a real sanitation problem in the town of Bambey[22] . The presence of uncontrolled refuse dumps raises health issues that the town council must address. The behaviour of local people in pouring wastewater into the streets or throwing household rubbish into the street is becoming a headache for the municipality, which is hampered by a lack of the financial resources needed to set up a system for collecting household rubbish. However, the UCG's intervention in the municipality has brought some relief to the local population.

The physical urbanisation of Bambey has not occurred at the same time as the sociological

[22] Reference Urban Development Plan for the town of Bambey

urbanisation of the population. The latter, still adopting a rural lifestyle, have a strong presence in the town and run small farms on the outskirts of the town.

The following graph shows the different factors in the urban dynamics of the town of Bambey according to the data collected from the households surveyed.

Figure 6: Survey opinions on factors affecting the city's urban dynamics

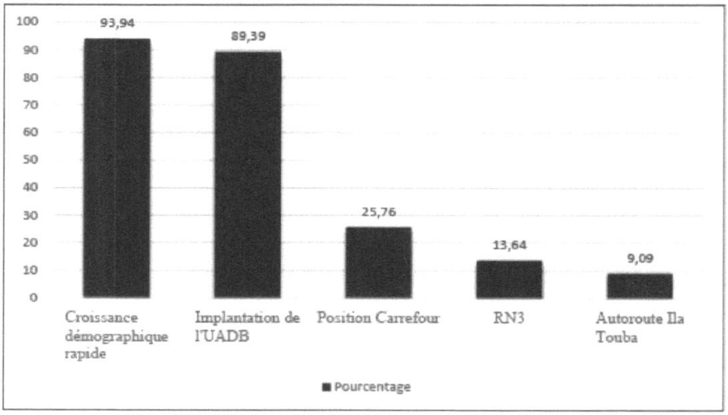

Source: Survey data, Guèye, October 2022

This graph highlights the different factors in the city's urban dynamics according to the surveys. In fact, population growth and the establishment of Alioune Diop University are the main factors in the town's spatial development. According to the responses obtained from the heads of households, 93.94% of the respondents stated that population growth was at the origin of Bambey's spatial dynamics. In the same vein, 89.39% of the responses from the surveys attributed the cause of the town's expansion to the establishment of the university. In addition to these two factors, the results of our survey also show that the town's crossroads location, its position along the RN3 and the recent Ila Touba motorway have all contributed to the town's attractiveness and hence its urban dynamism.

Figure 7: Survey opinions on the impact of the university on the city

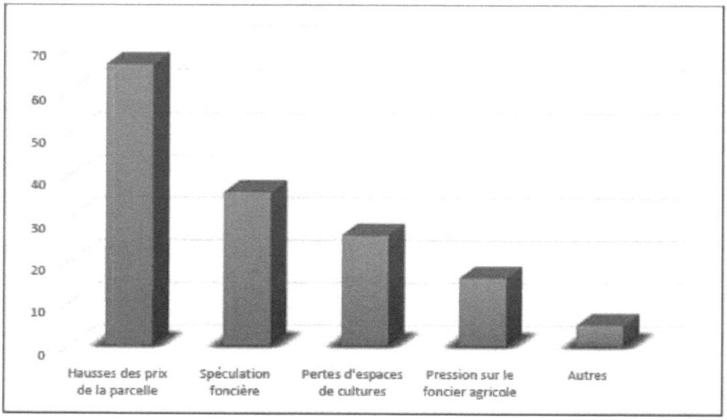

Source: Survey data, Guèye, October 2022

In the current urban dynamic of the town of Bambey, the university is the central nucleus around which several facts revolve. The establishment of the university in the town has given rise to a new urban centre at the entrance to campus 2, with the development of commercial activities creating a new market that operates on a daily basis.
According to our survey results, the university's impact on the city's economic development is undeniable. But we have to admit that the impact on land is much more remarkable. The main consequences are difficulties in accessing land, accentuated by the rise in the price of plots, land speculation, the loss of cultivation areas, and pressure on agricultural land, among others.
However, it seems much more important to look at the various problems associated with urban growth in the section that follows, because although cities have emerged as the engines of socio-economic development, they are also home to a host of issues and, above all, challenges: transport, resource management, the environment and public health (DIOP D. and TIMERA A. S., 2018).
In summary, the urban dynamic of the town of Bambey can be divided into three phases or periods. After experiencing linear urbanisation around the main communication routes during the colonial era, up until the 2000s, due to its role as a very dynamic stopover town in the trading economy with the marketing of groundnuts, Bambey saw its urbanisation process accelerate after 2004. However, even if this urbanisation was rapid, it was controlled, as all extensions were carried out within the initial boundaries of the town and obeyed well-defined urban planning rules. However, since the establishment of the Alioune Diop University in Bambey in 2004, the town has become much more attractive, and has become the preferred location for the surrounding rural population. This situation accentuated the occupation of the rest of the town's communal perimeter, leading to its exhaustion and plunging the town into a new dynamic with extensions in all directions. The main factors behind this urbanisation of the town are social, political and, above all, economic. The development of facilities in the city attracts people who, once in the city, find themselves in situations where the search for accommodation is sometimes very complicated. As a result, today's city, despite its exceptional development, is experiencing many problems due to its rapid urbanisation. So we're talking about the challenges of controlled urbanisation, despite the many issues surrounding land.

II PART

PERIURBAN LAND ISSUES AND DISPUTES

Bambey is a city that is growing rapidly, despite the many problems that the authorities are being called upon to tackle in order to make it a sustainable city and fully comply with the sustainable development objectives (ODD 11 Sustainable Cities and Communities). This is why this section first looks at the various challenges that the local authorities must meet in order to provide a concrete response to these urban problems, which are already having a major impact on the development of the town and the health of the population. Secondly, we analysed the land issues arising from the city's spatial dynamics.

Chapter I: Bambey, a fast-growing town facing new challenges

This chapter analyses the problems facing the city, the main factor being rampant, unplanned urbanisation. These include the development of precarious neighbourhoods and squatter settlements, problems of insalubrity, sanitation and, above all, the mobility of the city's populations.

I.1A failing sewage system

As in most towns, the problem of liquid and rainwater sanitation seems insoluble. Urban development, limited municipal resources, low population incomes and management problems are real obstacles, and they are very serious[23] .

Like some of Senegal's cities, which are characterised by a lack of planning and spatial organisation, the town of Bambey, despite its small size, may seem like an easy town to govern, but the reality on the ground is quite different. In fact, the town's urban development problems can be summed up as the absence of a sewage network, problematic household waste management causing waste water to stagnate even in the dry season, and the presence of uncontrolled rubbish dumps in all the town's districts. However, the city has two stormwater drainage and storage facilities (TINE M. L., 2015).

The rainwater and runoff drainage infrastructure consists of a drainage network 5,170 metres long. This network consists of two culverts, one of which is open and the other closed (TINE, M. L., 2015). In addition to this infrastructure, however, there is a recent 44,000-metre-long closed channel running in a north-south direction through the city before ending in the south-east of the DVF district, where its outlet is located.

Image 1: Open canal

Cliche : Gueye O., *October 2022*

This canal was dug in the Leona Nord district along the entire length of the road in order to free the area from rainwater. However, there is an open canal running through the same area. The construction

[23] El Jihad M. D., 16 March 2005. *Croissance urbaine etproblemes d'assainissement liquide etpluvial dans le bassin du Srou (Maroc central),* in Secheresse, n°1, vol. 16, 22 p.

of this open canal is proof enough of the critical state of the city's drainage system. It's a rainwater drainage system, but at the same time it's a public hazard because it's just a few metres from the town's rubbish tip. This raises real health issues, such as the development of mosquitoes and the spread of diseases like malaria, diarrhoea and bilharzia.

Image 2: Canal with closed sky

Cliche : Gueye O., *October 2022*

The lack of a sewage network and the installation of a rainwater drainage system in the city centre, crossing the Leona Sud and Nord neighbourhoods and DVF to the east, show the limitations of the city's sewage system. Despite this infrastructure to relieve the city of flooding, the situation today is even more dramatic. According to our survey results, most households are affected by flooding, with rainwater stagnating in the city's streets and even invading some of the houses in the Leona Nord and DVF neighbourhoods. This is due to the absence of secondary channels to drain the stagnant water in the heart of the neighbourhoods towards the main channel. According to the results of our survey, 80.3% of households surveyed are affected by flooding. To deal with this situation, people in certain neighbourhoods, such as Leona Nord, have dug an open canal along the road leading to Baba Garage to keep rainwater out of the area. The septic tanks that spill into the alleys, especially during the winter months, are a harbinger of excessive health risks, especially as children are constantly playing football next to these tanks.

The unsanitary state of the city's streets and the invasion of the canals by solid waste encourage flooding by saturation from below. The nature of the soil, with its high water retention rate, is one of the factors that encourages flooding. Mosquitoes are present throughout the year. The accumulation of rubbish and the proliferation of illegal rubbish dumps are indicators of the problem of waste management in the city. Although there is a system of household waste collection set up by the municipality with donkey carts, this system is still insufficient in the eyes of the population to rid the town of its rubbish. One person (GAYE S.) said: "*These carts can't do anything about collecting rubbish. The city is too crowded. Nothing; no effort from the town council. Which is deplorable, because in some alleys there's a nauseating smell. This year, mosquitoes have invaded our homes during the winter. Living in this town, especially in the rainy season, is a disaster with huge health risks*" (survey data, Gueye, October 2022).

The lack of a complete sewerage system throughout the city is not without consequences for the living environment and health of the population (TINE M. L., 2015). Clogged gutters are a fact of life in the city, especially in the Leona Nord district, and this is due to unsanitary conditions and the lack of an efficient system for collecting household waste. All of which encourages flooding. Here are some images taken as part of our investigations, showing the problems of sanitation in the city and the consequences for people's health.

These images show the presence of uncontrolled rubbish dumps in neighbourhoods and stagnant rainwater in the streets. All the necessary conditions are in place for the emergence of certain diseases in the city.

Image 3: Rubbish dump behind the bus station

Image 4: Rubbish on the street in the DVF district

Image 5: Flooding behind the bus station

Image 6: Open-air canal in the Léona Nord district

Image 7: Fly-tipping in the DVF district

Image 8: Stagnant rainwater in the DVF neighbourhood

Photo credit: Guèye O., *October 2022*

Despite this unfortunate situation, flooding problems seem to have lessened in recent years. According to the head of technical services at the town hall, "*flooding is still frequent in the town, but not like in 2005 and 2010. That's because the town of Bambey benefited from a sewerage system in 2012 from the State of Senegal as part of the ten-year flood control programme, to the tune of 1,937,000,000 CFA francs. As a result, the town is now breathing easier. However, during periods of heavy rain, such as August this year (2020), there are flooded areas in every neighbourhood. This is due to the geomorphological characteristics of the site"* (Survey data, Gueye, October 2022).

The presence of a large number of uncontrolled rubbish dumps in the city, in almost every district, and the lack of a sewage system mean that people tend to dump waste water in the streets. This creates public health problems, with the development of breeding grounds for diseases such as malaria and bilharzia.

Despite this worrying overall situation, there is still reason to put things into perspective. Not all neighbourhoods are affected to the same degree by the sanitation problem. In fact, the sanitation situation in the city conceals disparities between neighbourhoods. The Leona Nord and DVF neighbourhoods are the most affected by flooding, unlike the Leona Sud and Wakhaldiam neighbourhoods, which form the city centre and are less affected by flooding because there are no shanty towns and the rainwater drainage network crosses them in all directions.

Figure 8: Proportion of households affected by flooding according to the surveys

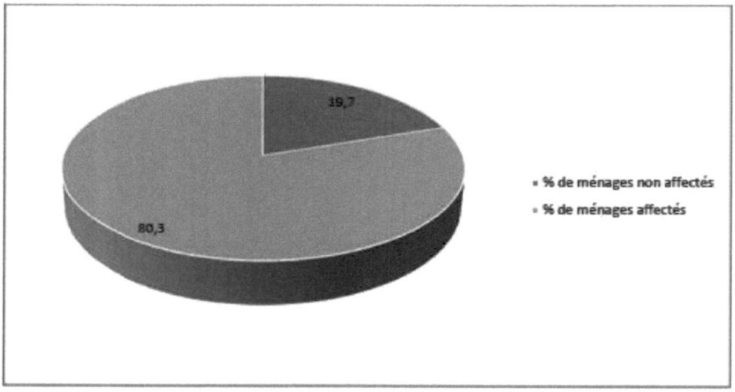

Source: Survey data, Gueye, October 2022

Waste management in the city is a real problem. The town has no landfill site, and the waste collection system in place remains highly unsatisfactory. All you have to do is drive around the town to see the rubbish dumped in every district.

The table below shows the households affected by household waste management in the different neighbourhoods.

Table 6 Proportion of households not affected by the waste management problem, by neighbourhood

Neighbourhoods	Proportion
DVF	32
Leona South	10
Leona Nord	43,75
Wakhaldiam	73,33

Source: Survey data, Gueye, October 2022

Table 7 Proportion of households affected by waste management problems, by neighbourhood

Neighbourhoods	Proportion
DVF	68
Leona Nord	90
Leona South	56,25
Wakhaldiam	26,67

Source: Survey data, Gueye, October 2022

These two tables show that the proportion of households facing household waste management problems differs from one district to another. The Leona Nord district is the one most affected by this problem. In fact, 90% of the households surveyed had to deal with household waste management. After Leona Nord, it was the DVF neighbourhood that came out on top. In fact, 68% of households surveyed in this district had to deal with waste management. The Wakhaldiam district, on the other hand, was less affected, with only 26.67%. The figure for Leona Sud was 56.25%.

In addition to the problems of sanitation in the city, the transport sector is also dysfunctional, with motorbikes and horse-drawn vehicles predominating.

I.2 Transport, a sector dominated by two-wheelers and horse-drawn vehicles

In African cities, the vast majority of which are undergoing rapid changes affecting the economic, political, social, cultural, environmental and health spheres[24] , urbanisation poses a particular challenge: the issue of mobility (NDIAYE E.M., 2013). Population flows in the country's secondary towns pose a real problem. Following the example of the town of Bambey, the lack of transport means that people now get around the town on foot or by motorbike. The lack of bus services in the town and of taxis to move people and their goods around the town are a real bottleneck in terms of mobility.

In the town of Bambey, walking is the means of transport for 33% of households surveyed. However, 32% use motorbikes to get around, and 24% use carts. Cars were used by 11%, making them the least used means of transport. This can be explained by the absence of a bus service, and the very small number of taxis, or even none at all, as most of the cars in the town provide intercity transport between Bambey and Senegal's capital, for example. In terms of mobility and transport, Bambey is nothing more than a transit area, a kind of relay town between Touba, Diourbel and the country's capital. Here, in the context of our study, the mobility in question is that which focuses on the movement of people within the urban perimeter of the town of Bambey. Thus, interurban travel or the flow of people between the city and rural areas is excluded, as is the flow of people between Bambey and the rest of the country. As a result, means of transport for long distances outside Bambey are not taken into account; only flows within the city are considered.

Image 9: Garage for motorbike taxis

Cliche : Gueye O., *October 2022*

Today, the city's development, with its extensions on all sides, with key points such as the university to the west and the centre for Islamic studies to the east, as well as the concentration of service facilities in the city centre, requires an efficient public transport system to be put in place.

Mobile needs will continue to grow and the population will continue to increase. To address this shortcoming, the new dynamic on which urban policies must be based in the coming years, transport must play a key role. Transport infrastructure and mobility play an important role in defining the city of tomorrow. "*They have to keep pace with the emergence of new technologies, the diversification of transport modes and the revolution in transport practices*"[25] .

The following graph shows the proportion of means of transport used by the population in the city

[24] Sarr I. (2006), Cited by Ndiaye E.M., 2013

[25] Brueder P. et al, 2022. *Quelle mobilite pour la ville de demain? Un regard croise des experts et de la litterature scientifique,* in open ёйюп, volume 22, n°1

according to the surveys.

Figure 9 Means of transport used by the population according to the surveys conducted

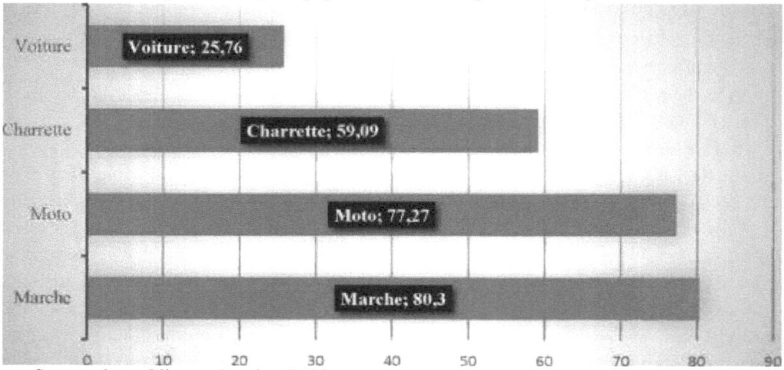

Source: Survey data, Giieven October 2022

I.3 Development of precarious housing on the outskirts of the city

Lack of access to decent, secure and affordable land is the main reason why there are so many informal settlements in African cities, and it is also a contributing factor to urban poverty (UN-Habitat, 2011). The issue of irregularity in the city is a recurring theme in urban planning, for which rules and standards are the main tools. These rules are imposed on the population by regulatory urban planning or directly applied in operational urban planning projects[26] . As Le Bris et al (1983) state, it is not enough to have theoretically perfect laws; the government must also be able to apply them on the ground. However, urban and peri-urban land management is hampered by a number of factors. According to LE BRIS E. et al (1983), "there is a *need (...) to distinguish, on the one hand, between usage behaviour and, on the other hand, creative behaviour. It is the latter that is usually the focus of analysis of the phenomenon: shanty towns and spontaneous urbanisation are the focus of attention. The official city is contrasted with a sort of "bisected city" created by the population to meet unmet housing needs.*

Today's world is becoming urban, and African countries, like Senegalese cities, are subject to the reality of massive urbanisation. This is leading to the anarchic growth of cities and the multiplication of their urban extensions, which are generally described as informal MICHEL M.R., DIOP D. et al (2013).

One of the urban phenomena on the outskirts of Bambey is the proliferation of irregular neighbourhoods. The massive influx of rural dwellers into the town is having a huge impact on the harmonious development of the town and its spatial organisation. The development of irregular or spontaneous neighbourhoods to the south of the town towards the commune of Ngoye, but also to the east towards the commune of Thiakhar and to the north of the town, is compromising the control of its spatial development. In fact, the settlement of these populations without prior planning and spatial organisation gives rise to a situation that encourages insalubrity, flooding and many other phenomena of this kind.

Image 10: Precarious housing in the south of the DVF district

[26] Clerc-Huybrechts Valerie, 2002. *Les principes d'action de l'urbanisme, Le projet Elyssar face aux quartiers irreguliers de Beyrouth*, doctoral thesis in urban planning, Universite Paris 8.

Cliche : Gueye O., *October 2022*

This image is a perfect illustration of the existence of precarious housing on the outskirts of the city. This is a group of houses located to the south of the DVF neighbourhood on the way to the village of Battal in the commune of Ngoye.

Today, with the expansion of the city towards its outskirts, we are witnessing the development of precarious neighbourhoods which, in addition to being characteristically poor, are a sign of uncontrolled urbanisation with its huge impact on land and the environment. As a result, a number of issues have arisen around land on the outskirts of the city.

Chapter II: Land issues arising from excessive urban growth in Bambey

As everywhere in the world, talking about land issues following the advance of the city or its spatial expansion gives rise to unprecedented effects or impacts, so that we often talk about issues. These land issues, which take shape and form with urban development, are manifold and begin in this case with what is known as land security. In fact, with the spread of the town of Bambey and the threats it poses to agricultural land in the neighbouring communes, in the face of the decree extending the territorial boundaries of the commune of Bambey, which is very much contested by these territorial collectivities, the former rural communities, each municipality is doing its utmost to protect and secure its land base. So we can see that the urban phenomenon of spatial growth in towns and cities is giving rise to many different land issues. This poses enormous difficulties for decision-makers and development players in terms of access to housing and basic social services. The spatial dynamics of the city of Bambey are reflected in the inclusion of periurban areas, sites of transition between the city and the countryside (MBAYE I. et al. 2018).

But beyond this desire to secure land, there is something else lurking, all the more so as the law on the national domain has already ruled on land management methods, even if its application poses problems in many cases. This suggests that securing or protecting land in the face of urban development raises political, economic, socio-cultural and environmental issues. But before analysing the land issues raised by urban dynamics, we have attempted to look at the evolution of land governance in Senegal, the conditions of access to land and the land allocation process.

II.1 Land governance in Senegal

Characterised by the coexistence of colonial and customary law (CLAVEL M. 2018), Senegal's land tenure system has long been a source of much incomprehension, as the influence of modern law is largely confined to the major cities.

In 1964, the National Domain Act abolished customary rights in order to secure land tenure. However, it is clear that despite the existence of such a law, it is proving complicated to convince a peasant inheriting a plot of land from his ancestors that it does not belong to him, but rather to the State. However, Article 1 of the National Domain Act of 17 June 1964 stipulates that: "*All land not classified as public domain, not registered and whose ownership has not been transcribed at the Registry of Mortgages as of the date of entry into force of the present Act, constitutes the national domain as of right. Nor does land which, on that same date, is the subject of a registration procedure in the name of a person other than the State, form part of the national domain as of right*".

The 1964 law on the national domain does not make it impossible for people to access land. In fact, according to decree no. 72-1288 of 27 October 1972 on the conditions for allocating and disallocating land in the national domain included in rural communities, amended by decree no. 80-445 of 10 April 1986, article 2 states: "*Land for cultivation and clearing is allocated by a decision of the rural council*". However, this resolution is not enforceable until it has been approved by the departmental prefect (Article 24 of law no. 72-25 of 19 March 1972 on rural communities). By entrusting local or territorial authorities with the management, organisation and planning of their development, the State has promised them autonomy and independence (FAYE J., 2008 cited by CLAVEL M., 2018). All of its laws and decrees relating to land management in the country, which are increasingly tending to make local authorities responsible for land management, mark the beginnings of decentralised land management. As part of its decentralisation policy, the State is making use of these opportunities to transfer the land required for expansion to the communes, and to proceed directly with developments intended for urban housing (FAYE J., 2010).

This decentralisation of land management is reflected in the transfer of land-related powers under Act II of decentralisation. Nine areas of responsibility were transferred to local authorities, including land. Local authorities are expected to draw up urban development plans, as well as planning their development over time and space. They therefore have their destiny in their own hands in order to promote harmonious local development, even if others see this as a disengagement on the part of the State. This is something that cannot be denied, given the urgency of the context for local development.

In 2013, Act III of decentralisation radically changed the country's political and administrative organisation, yet without calling into question the administrative division of local authorities, which was often a source of conflict and dysfunction (CLAVEL M., 2018). Today, in the face of obstacles to the implementation of intercommunality between several communes, we are witnessing conflicts or disputes linked to the blurred administrative boundaries of local authorities. This is the case between the commune of Bambey and its surrounding rural communes, because the spatial development of the commune of Bambey means that the town now needs more space to contain its population. But this is only possible when the boundaries of the commune are pushed back on all sides. On the other hand, given the many issues that may arise from the urban sprawl of the town of Bambey onto the agricultural land of the surrounding communes, this fact remains difficult to achieve, especially as it comes up against fierce opposition.

11.1.1 Access to land in Senegal

In Senegal, access to land is a serious issue that deserves to be addressed with a great deal of attention and flexibility. In Senegal's constitution, it is true that the principle of equality in terms of access to land, recognising the same rights for women as for men, is still lacking for women. But in this context, we are interested in outlining the different procedures to be followed to gain access to land. There are three types of land in Senegal: land in the national domain governed by law 64-66 of 17 June 1964 on the national domain, land in the State domain governed by law 76-66 of 2 July 1966 on the State domain, and land in the private domain governed by law 2011-07 of 30 March 2011 on land ownership. As a reminder, article 15 paragraph 2 of the 2001 Constitution of Senegal states: "... *men and women have an equal right to possession and ownership of land under the conditions determined by law...*". However, there is a procedure to follow for access to land. This procedure varies from case to case. Whether 9a land of the national domain, land of the area of the terroirs or land of the classified zones.

11.1.2 Ground access procedures

For access to land in the national domain, the 1964 law on the national domain, which also governs agricultural land, rules out any idea of gender discrimination in access to land. For land allocation, which is the process of making national estate land available by municipal resolution, the main principles guiding land allocation are the following Land allocation is guided by a number of key principles.

They are :

I am a member of the locality (individual member or several members in an association or cooperative);

make the most of the plot;

enjoying a simple right of use.

In addition to these three criteria, however, it should be noted that assignment confers only a right of use and is indefinite. However, it may be withdrawn at any time under the conditions laid down by law. This procedure is reserved exclusively for land in the national domain.

For land for residential use that can be accessed following a subdivision, which under the terms of article 14 of the French Town Planning Code "*Constitutes a subdivision, the operation having the effect of developing, equipping and dividing into lots one or more pieces of land for sale or transfer free of charge or for rent*".

From the application to the closing of the land file, via the decision of the municipality and the physical installation of the applicant, the land allocation process follows a whole procedure summarised in the diagram below.

Figure 10: Diagram of the land use process

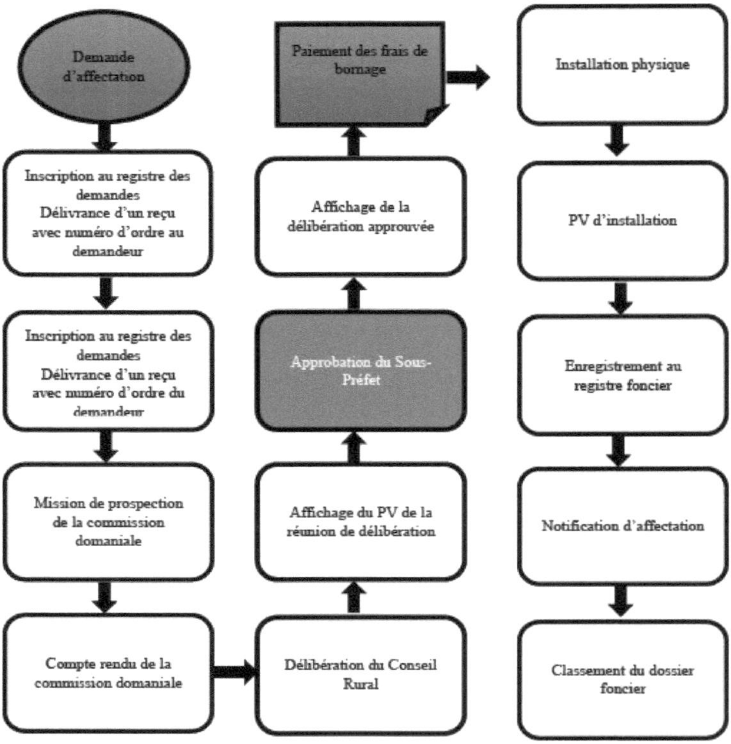

II.2 The land issues involved in sustained urban growth in Bambey

Nowadays, most towns have reached their territorial limits, and their surface area is expanding into the land of the surrounding communes, with the development of housing on the outskirts of the town. Thus, as land is seen not only as a factor of production but also as a tradable commodity, it is becoming a major issue that cuts across different administrative districts, each of which is seeking to defend its own interests. This is the situation between the commune of Bambey and the communes of Ngoye, Thiakhar, Ngogom and Ndangalma.

In the town of Bambey, talking about land issues means discussing the difficulties encountered by the authorities in controlling the town's spatial development. For years now, the town has been expanding onto the agricultural land of other communes, and the authorities are finding it difficult to find a solution to this problem in order to avoid any land disputes with the local political authorities of the other communes.

The land issues raised by the city's urban growth come down to access to land for housing, infrastructure construction and facilities for the city authorities. For the communes of Ngoye, Thiakhar, Ngogom and Ndangalma, the main issues are preserving agricultural land and avoiding the loss of demarcation fees, a major source of financial revenue for the municipalities. All the initiatives put in place by these former rural communities to secure the land within their communal boundaries are guided by the need to preserve the economic interests of their municipalities.

By summarising the expression "land issues" as what can be gained or lost from land, we can clearly understand the many and varied issues and debates surrounding land between the commune of Bambey and the other surrounding communes.

What follows is an analysis of the political, social, cultural, economic and environmental land issues related to the spatial dynamics of the town of Bambey.

II.2.1 Political and socio-cultural issues

Immediately after independence, Senegal's political authorities demonstrated a desire to democratise access to land with law no. 64-46 of 17 June 1964, which defined the Domaine National. This initiative implied a genuine social change that went beyond legal issues (IPAR, 2010)[27] . From then on, a logical opposition arose, with the confrontation on the ground of two conceptions of land and its use. Recent initiatives to revive agriculture are reinforcing the debate on land redistribution and, consequently, on the laws governing land management.

The dualities between the town of Bambey and the other communes show that the process of decentralising land management is not satisfactory and that the effective application of the principles and guidelines of the law on the National Estate and its implementing decrees still needs to be reviewed. In terms of land management, human, financial, material, institutional and even structural constraints or obstacles hamper the actions of the executive and decision-making bodies of local authorities. As with other transferred powers, these are characterised by a lack of resources and little power to exercise them (IPAR, 2010).

Thus, the willingness of local authorities to protect their land is a shield to protect agriculture on the outskirts of the town where the people live, since these communes are inhabited by peasant farmers, and agriculture and livestock farming are the basic sectors of their economy. So the preservation of family farms and the reduction of poverty in rural areas are among the reasons why the municipal authorities of the other communes are opposed to any attempt to include their land in the Bambey land register. In reality, however, the argument put forward by the local authorities of the other communes may be quite the opposite. Awareness shows that sooner or later, the agricultural land around the town of Bambey will change use to the detriment of farming. So the will to keep one's land and receive taxes on it, even if they have not yet been applied in full and rigorously in most towns in Senegal, is a good thing. Because leaving this land to the detriment of this activity and to the benefit of urbanisation would deprive these communes of an important source of funding such as demarcation fees and local land taxes, which will now be paid by the commune of Bambey. It is therefore easy to understand why the local authorities in the former rural communities are having such difficulty in seeing the town enjoy the advantages conferred on it by its decree extending the territorial limits of the commune.

II.2.2 Economic and environmental issues

The issue of land security raises different conceptions of land use and management. According to the IPAR report (2010), two trends can be identified. The first is the idea that land privatisation is a priority solution for securing investment, and is a guarantee of access to credit and the development of entrepreneurial agriculture. This is the view of the Bretton Woods institutions (WB and IMF). The second is simply to recognise local rights, which are in fact flexible, dynamic and likely to lead to a type of security favourable to investment and the development of family farming. But what both have in common is the securitisation of land, it's just the way in which it is done that differs. In August 2020, due to the lack of arable land and people's concerns about the town of Bambey's subdivision plans, the people of the commune of Ngoye voiced their disagreement with this land-consuming project, which is dangerous for the families who have most of their fields in the area. The subdivision of this land would also have economic and environmental consequences, with the destruction of the plant cover, the felling of trees, the disappearance or reduction of grazing areas and the loss of biological diversity, making it difficult to carry out any agro-pastoral activities.

On the other hand, Senegal's commitments to achieving the SDGs will remain under threat if such practices continue to be observed throughout the country. In particular, MDG 16, "*Sustainable cities and communities*", should guide the actions of local and regional authorities in terms of urban development and planning, because the climate context in which we live does not encourage such

[27] Initiative Prospective Agricole et Rurale, Gestion fonciere au Senegal : enjeux, etat des lieux et debats, 2010

practices.

Chapter III: Land disputes in the town of Bambey

Population growth and the development of infrastructure and facilities justify the city's attractiveness, and therefore its sustained urban growth. As a result, this attractiveness means a year-on-year population surplus, which in turn creates a need for housing, and therefore new land for development. However, the city has reached its territorial limits on all sides. Strategies have been initiated by the state and local authorities to meet its housing needs and build facilities for the city's population. However, these initiatives on both sides have led to land disputes. Some of these are linked to administrative divisions, while others stem from customary practices or the domination of customary law over legal law in the area.

In this section, we have attempted to analyse the various land conflicts in the city. To do this, we have divided these conflicts into two categories, according to their nature and the actors involved.

III.1 Urban land management: between political will and reality on the ground

In the immediate aftermath of independence, when Senegal introduced a law to secure land, it was thought that the problem of land management would be forgotten. But it has taken more than 50 years today for us to realise that in reality the problem of land management remains as topical today as ever. Indeed, attempts at land reform seem to have resulted in inadequate measures if we look at what is happening on the ground on a daily basis. The issue of good land governance in Senegal in order to eradicate the problems of insecurity and land corruption was part of the various phases of the land reform process in Senegal.

The spatial development of towns and cities always has spatial, political and economic consequences. In most cases, these consequences relate to land, with the risk that the initial territorial limits of towns and cities will be overstepped and agricultural land consumed. This often leads to land disputes between territorial communities or local political authorities. In fact, *"the appropriation of land is carried out on the fringes of law no. 64-66 of 17 June 1964, which is intended, among other things, to standardise and regulate local land tenure. Peripheral land tenure is a source of social upheaval, especially at the level of customary ownership. If it's not occupation of State reserves, which some consider to be "vacant land" with no owner, it's "wild" and "malicious" speculation on marginal land. The result of either form of land production and acquisition inevitably leads to spatial disorder and urban anarchy"* (WADE C.S., 1995).

While it is clear that the problem of land in Senegal has been resolved to some extent by the law on the national domain and its implementing decrees, the State is now constrained in certain projects by customary law, which also makes its laws in the management of the country's land. Today, the State is constrained in certain projects by customary law, which also sets the rules for land management in the country. The State does not dare

not apply the provisions of the law on the national domain to the full. Indeed, even if he declares that all unregistered land belongs to the State and is classified in the national domain, he is not unaware of the sanctity of customary land legislation, which he cannot afford to ignore.

The State and local authorities often come up with mechanisms for managing land without a track record, because local authority land is State land.

Land disputes between local authorities are becoming increasingly common in Senegal. According to the Ministry of the Interior's 2020 national directory of land disputes, 307 land disputes have been recorded, affecting all regions but of varying natures. While some are caused by village or commune boundary issues (19.86%), others are caused by disputes between private individuals (34.20%), or between local people and developers (13.02%), while community disputes (14%) and subdivision issues (18.89%) are marked by disparities between regions, such as Dakar, with 9 disputes caused by subdivision issues, 5 in the Saint-Louis region and 4 in the Diourbel region.

The problem of the delimitation of communes is becoming more and more widespread these days, and these types of conflicts have not been seen in the country for a long time. This is due, on the one hand, to the development of communes as departmental capitals and secondary towns, which, in their recent dynamic, are gaining more and more ground and spreading out towards neighbouring communes. And secondly, the absence of a rural land register and a well-maintained land register in most rural communes.

III.2 Land disputes in the town of Bambey

The groundnut basin has become a hotbed of land disputes. According to a report by the Ministry of the Interior (Cartographie des litiges fonciers et conflits communautaires au Sénégal, 2022), the Diourbel region alone has more than a dozen conflicts. Some of these are due to boundary issues between communes or villages, while others have their origins in subdivision issues and others in conflicts between private individuals. In the town of Bambey, depending on the actors involved, we have identified two types of land dispute, detailed below.

Most of Africa's medium-sized towns were created for political, administrative and commercial purposes. In fact, the economic boom in these medium-sized centres rapidly led to a massive influx of rural dwellers in search of better living conditions.

This massive influx of new people is reflected in extensive urban growth. The surplus of births is added to the population movements that have made a major contribution to the current urban dynamics of second-tier towns. These urban centres are becoming attractive, and we are seeing large numbers of migrants from all over the world. This population and housing dynamic means that medium-sized towns need to find new spaces to satisfy the high demand for housing plots. This massive demand for housing and dwellings is now one of the causes of many land conflicts in medium-sized towns (MBAYE I. et al., 2018), following the example of the town of Bambey.

"*A land dispute is a disagreement between two or more parties over a piece of land*" (MBAYE I., 2018). The population arriving in urban areas requires the construction of housing, which contributes to urban sprawl. Consequently, housing remains the main means of spatial expansion in Senegalese cities, especially medium-sized cities (MBAYE I. et al., 2018). The scale of the city's demographic growth has been accompanied by exponential urban dynamics.

Its rapid urbanisation raises the issue of land conflicts between the town and its neighbouring communes. Excessive urbanisation thus poses acute problems, particularly access to land (MBAYE I. et al., 2018). "*Land has thus become an object of covetousness and exchange, but also a source of conflict. Because many urban conflicts stem from land*" (MBAYE I. et al., 2018). "*Land has become a valued economic object and is being transformed into a commodity. This phenomenon is reinforced in cities by considerable demographic growth*" (Lavigne-Delville, 2000, cited by MBAYE I. et al., 2018).

III.2.1 Land disputes between local authorities as a result of administrative division

For several years now, the commune of Bambey has been in a tug-of-war with the communes of Ngoye to the south, Thiakhar to the east, Ndangalma to the west and Ngogom to the north. It all started with the desire of the municipal authorities of the communes of Bambey to subdivide land belonging to these communes following Decree No. 2011-1991 of 16 December 2011 amending and extending the territorial boundaries of the commune of Bambey. This decree caused a split between the commune of Bambey and its four neighbouring communes[28] . But this dispute between local authorities over the decree and its proposed extensions cannot be reduced to a simple disagreement over the application of the decree and its consequences. The sites of the CNRA and the Alioune Diop University in Bambey, with areas of more than 30 hectares, belonged to the rural communities of Ngogom for part of the CNRA and Ndangalma for the University before being transferred to the land of the town of Bambey. As a result, all the land revenues generated by these structures are gobbled up exclusively by the town of Bambey. This is a source of frustration for these two government bodies.

The master plan for urban development up to 2021 provides for new extensions to the east, north and mainly south of the town, as the CNRA wall in the north now seems to be an obstacle to any plans to

[28] Journal lequotidien, Diourbel-Litige foncier a Bambey : Des habitants du Baol se crepent le chignon, in lequotidien.sn, 14 aout 2019

extend the town to the north. This PDU, drawn up on the basis of the decree extending the municipal boundaries of the town of Bambey, has been attacked by the municipalities of Ngoye, Ngogom, Thiakhar and Ndangalma, who have formed a united front against the decree. Because it has not been published on the official gazette website, these former rural communities, which are moreover more affected by the decree and have had large areas of land expropriated, with all the attendant consequences, are vetoing the application of the decree following attempts by the Bambey authorities to subdivide land in the communes of Ngoye, Thiakhar and Ndangalma, as well as in the commune of Ngogom.

These local authorities, aware of the land revenues they would lose to the commune of Bambey, could not remain idle in the face of such an attempt. So they challenged the decree extending the boundaries of the town of Bambey before the supreme court. According to the head of Bambey's town planning and housing department, the regional governor asked that nothing be done on the land until the Supreme Court had ruled on the case. However, it has to be said that the decree has not been published in the official gazette and therefore cannot be applied. But in this case, according to the head of the Bambey town planning and housing department, the key point is the lack of information on the part of the municipal authorities of these recalcitrant communes, because the subdivision projects they were contesting were not carried out on their land base, but were in fact subdivision projects that should have been carried out and were carried out on land belonging to the town of Bambey. These were two housing development projects involving the extension of the DVF and Leona Nord neighbourhoods. This development, which covered an area of 82 hectares in the Leona Nord district and 52 hectares in the DVF district to the south of the town, towards the commune of Ngoye, was just the answer to the many requests made year after year by the Bambey town council due to population growth and demand for plots of land.

Almost all of these four communes, which are crying out that their land has been sold off by the commune of Bambey, are essentially former rural communities, now very poor communes, which, in addition to the endowment funds received, have no other financial resources for the development of their municipalities. To deprive them of their land, and consequently of demarcation fees and other land revenues, would be to drive them to the brink of extinction, as they would not even have the funds needed to develop their land. So, to avoid losing these and other demarcation fees, communes such as Ngoye and Thiakhar have already begun projects to subdivide their land on the outskirts of the town of Bambey.

Today, given the exhaustion of the town's communal boundaries and the extension of the town onto land in the national domain belonging to the former rural communities, the question that arises is who owns the land on which the town is spreading out. While the town of Bambey claims to be the owner of this land, citing decree no. 2011-1991 of 16 December 2011 on the modification and extension of the territorial boundaries of the town of Bambey, the communes of Ngoye, Thiakhar and Ndangalma, which are mainly affected by the town's subdivision projects, are referring to the idea of land grabbing by the authorities of the town of Bambey.

The various subdivision projects initiated by the town of Bambey have met with endless opposition from the people of these communes bordering the town of Bameby. The most recent subdivision project, in June 2020, was fiercely opposed by the people of these former rural communities. In fact, on 1er August 2020, the people of the commune of Ngoye, the main commune to lose its land to the town of Bambey, organised a demonstration to denounce the parcelling out of their agricultural land by the town of Bambey.

The municipal authorities of the four communes of Ngoye, Thiakhar, Ndangalma and Ngogom have unanimously attacked the decree extending the territorial limits of the town of Ngoye.

Bambey. According to the municipal secretary of the commune of Ndangalma, this decree is not enforceable against third parties because it has not been published in the official gazette[29] . However, a

[29] Lequotidien.sn, 14 August 2019. *Land dispute in Bambey: Residents of the Baol dig in their heels*

decree is only enforceable after publication in the official gazette.
However, faced with this attempt to bring this land reserve within the city's boundaries, the communes of Ngoye and Thiakhar, which are more affected by the expansion of the city of Bambey, have retaliated by each launching a subdivision project to counter any new attempt by Bambey to reclaim their land. These legal restrictions on the expansion of the town of Bambey have prompted the head of municipal technical services at Bambey town hall to say that, after launching the subdivision project covering more than 20 hectares to the south and east of the town, the rural communes made it very difficult for us to block the subdivision project. They attacked us in the courts, and the communes of Ngoye and Thiakhar initiated subdivision procedures. This shows their jealousy of the town's development and their unwillingness to pay the demarcation fees, even though they don't have the means to develop the land.
Today, despite the fact that this situation is half resolved following the regional governor's decision to do nothing in these parts while awaiting the decision of the supreme court following the other communes' complaint against the commune of Bambey, it leaves an insecurity of land ownership that does not say its name. In fact, the commune of Thiakhar has already started to sell its land around the town of Bambey in order to counter any advance by Bambey in this area. This sometimes leads to conflict, as there is no shortage of confusion over land ownership[30] .

Map 9: Bambey and its four neighbouring communes

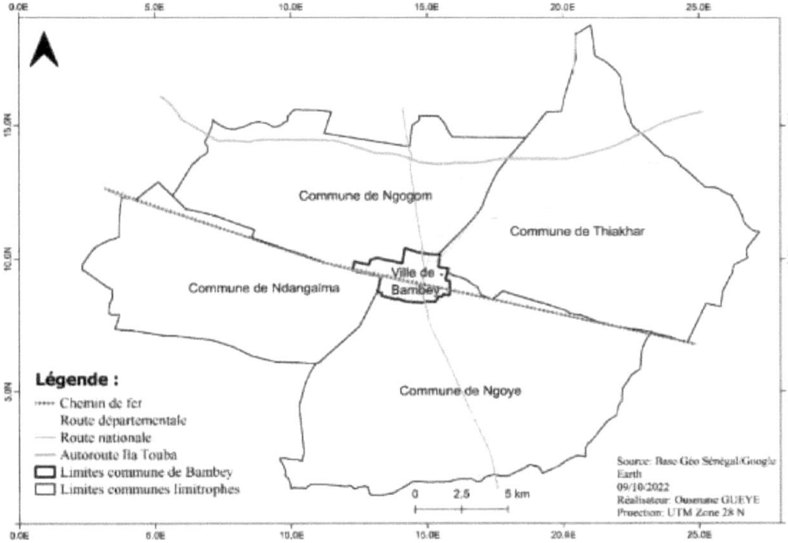

Source: Base Geo Senegal, Google Earth/QGIS

III.2.2 Land conflicts linked to customary practices

In addition to these problems between local authorities, the town of Bambey has been facing a problem of governance and land insecurity for the past three years. In 2020, the subdivisions initiated by the town council led to popular uprisings in the town, mainly in the DVF-Diamaguene and Leona neighbourhoods. In these two neighbourhoods, customary land practices are still a reality on the ground. In fact, before the housing development projects in 2020, people had already settled on the outskirts of the DVF neighbourhoods in the south of the city and Leona North. But these settlements

[30] Head of the technical department at Bambey town hall (interview on 23 August 2022)

had no legal basis.
In the DVF neighbourhood, the "baay deuk" phenomenon ([31]) meant that some people had already moved in before the development. In 2020, with the extension of the DVF and Leona Nord neighbourhoods, the housing development project had divided up their living areas, and some of them, after deliberations by the town hall, had to leave their plots in favour of other families. But even though the families living in this part of the town are in a housing area that has no town planning regulations or legal basis, these people are saying "no" to any idea of expropriating their land, because they don't intend to leave their usual homes, even though they have no town planning regulations. And yet, after noting that these people had developed the land they occupied, the local council decided to compensate them in kind. But the local people were opposed to any attempt to leave, because on the basis of customary law, they appropriated these areas and left their mark on them. This has caused a stir in the town, giving rise to movements to defend the land of Bambey.
This affair was beginning to take a different turn. It is becoming an issue between the municipality and other political leaders who claim to be playing a role in protecting the land of these diminished populations.
In 2020, after the subdivisions had been made, families who had already developed land but without any legal rules found themselves in an alarming situation because they were being expropriated from their plots. So, according to a member of the Collective for the Defence of Bambey Land, 27 families should be evicted from their homes and others should give up their homes to other people who, according to the Collective for the Defence of Bambey Land, are close to the town's mayor. But beyond this, land management in the commune poses a real problem, as Ndiaye S. tells us: "*We found more than 42 duplicates after the lists were published. There are also minors to whom the municipal team has allocated plots, and people living outside Bambey have benefited from these allocations, which is against the law"* (interview, October 2022).
This led to a complaint against Bambey town hall by the collectif pour la défense des terres de Bambey (collective for the defence of Bambey land) to the cours des comptes, the IGE and the OFNAC.
Today, the commune of Bambey is one of the communes in the department most affected by land disputes. In fact, after analysing the results of our survey, we found that 16.67% of respondents had encountered land disputes in recent years. However, this proportion varies from neighbourhood to neighbourhood. The DVF district is the most affected, followed by the Leona Nord district. This is justified by the extension of the city towards these two neighbourhoods with the latest housing development projects carried out in 2020 but also the arrival of populations from the surrounding countryside who, once in the city, settle in the pëH-pЬ-ёпез because land is much more accessible there but also finding the urban centre densely peиplё. According to the results of our survey, we can also note that land problems in the city are recent as more than 63.63% were notës in the city less than 5 years ago. On the other hand, 27.27% of conflicts occurred between 5 and 10 years ago, compared with 9.10% that occurred more than 10 years ago (DoniK'es d'enquete, Gueve 2022).
More than 36.37% of land disputes in the town are between the town council and the local population. However, 54.54% of land disputes are between the town's inhabitants. This is due to confusion over plots or problems Hё8 a Heritage in most cases. Land disputes between a family and a private owner account for 9.09% of land conflicts in the town of Bambey.

Image 11: Demonstration by the collective for the defence of land in Bambey in August 2020

[31] A Wolof expression that refers to the settlement of people on sites that are usually on the outskirts of towns, but which have no legal basis, exposing the inhabitants to land insecurity.

Source: Ndiaye S., *August 2020*

Figure 11: Rëpartition of land disputes by neighbourhood according to enquëtēs.

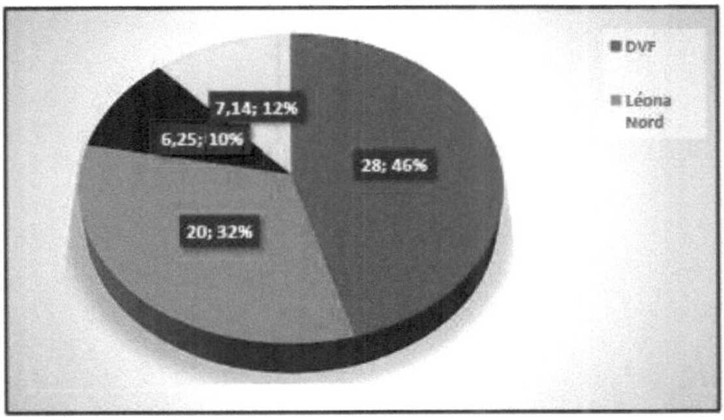

Source: Donees d'enquete, Gueye, October 2022

III PART

IMPACT OF URBAN SPRAWL ON LAND USE PERIURBAN AND DEVELOPMENT PROSPECTS

Chapter I: Urban sprawl and land speculation in Bambey

This chapter analyses the different impacts of urban spatial expansion on land. Firstly, we focus on urban sprawl and its consequences for periurban land. Secondly, we study land speculation and the growth of the real estate sector in the town of Bambey, before concluding with a discussion of the town's development prospects.

I.1 Peri-urban land under pressure from galloping urbanisation

Since their creation, cities have evolved over time and space until they have outgrown their natural sites, and this can be described as urban growth or urban sprawl. The latter refers to "*an increase in the surface area of an urban area greater than that of its population*" (BURGEL G., 2017). Urban sprawl is based on discontinuity and the over-consumption of resources, which generates huge environmental costs. According to (BURGEL G., 2017), urban sprawl is "*associated with excessive consumption of agricultural land, natural spaces and energy, as well as being a source of increased pollution and wasted time due to the need to travel increasingly long distances, mainly by car*".

Urban sprawl is a global phenomenon that varies greatly, but it corresponds to a logic that deserves to be explained. The expansion of cities began with the industrial revolutions, especially in Europe, but it was not until the aftermath of the Second World War that large human communities emerged and the process of urbanisation accelerated. But it is important to remember that the mutation and dynamics of urban areas are not exclusively quantitative and spatial. It is accompanied by profound social changes and a growing complexity in the notion of the city (PAULET J. P., 2009).

This urban expansion of the town of Bambey poses the problem of town-country relations in different terms, as the interpenetration of fields, wasteland and built-up areas is becoming the rule. The city walls have long since disappeared, and urbanisation has spread far beyond the inner city. As businesses moved to the outskirts of the city on the main roads, it became easier to create suburbs in the middle of the countryside on undeveloped farmland, with a market town or village at the centre of the new urban area. The former farming population will therefore be drowned in an urban type (PAULET J. P., 2009).

According to PAULET J. P. (2009), an estimated 476,000 hectares of arable land are transformed every year in developing countries. Agricultural land is moving away from towns and cities as a result of the strong land pressure exerted by the demographic weight of towns and cities. In more general terms, if we include rural areas, over a century more than 10,000,000 square kilometres of land will be degraded. The need for facilities and housing plots is often legitimate. However, they are very often built in conditions that are dangerous for the environment and for farming.

According to PAULET J. P. (2006), before the 20^{e} century, urban growth was relatively slow. But recently, according to MBAYE I. et al. (2018), "*the sustained trend towards urbanisation has accelerated at a worrying rate, resulting in demographic growth, densification of the urban fabric and unprecedented spatial urban sprawl, generating demand for land to meet the needs of the city. As a result, local forests, wetlands, agricultural areas and classified forests are falling victim to this process, despite the existence of a substantial legal arsenal. Cohabitation between the city and classified forest areas is becoming difficult. The city is expanding at the expense of these areas, which are seen as mere land reserves*". Thus, the demographic boom is transforming the contours of the town of Bambey by bringing certain parts of the countryside into the town area. The territorial boundaries between the town and the surrounding countryside have become very difficult to identify.

In Bambey and the surrounding area, urban sprawl is having serious consequences for the environment. These consequences are most notable in the disruption of the ecosystem, the loss of agricultural land, and the decline in arable land and agricultural yields. In this context marked by globalisation and economic liberalism, "*economic and even political rivalries put ecological interests in second place. The loss of agricultural land on a global scale is therefore the greatest danger*" (PAULET J. P., 2009).

The high price of land in the urban centre and the need for housing space are driving individuals and businesses to settle on the outskirts of the town of Bambey.

Between 1988 and 2002, the population of the town of Bambey grew significantly, with an additional population of 4,098 in 2002. In 2013, the additional population of the town was 8,542, i.e. double the additional population in 2002. This urban population growth is the basis for the estimated land requirements for extensions to the town, as the current situation in the town is that its communal perimeter is exhausted, with the exception of a few hectares available to the south of the town in the DVF district and to the north and north-east in the Leona Nord district, including the agricultural fields of the people of Keur Sakoura.

The town of Bambey is currently experiencing extreme land scarcity, as all its land has been subdivided by the municipality. With the exception of the 2014 subdivision, all the other subdivision projects in the town of Bambey are planned for agricultural land in the surrounding villages. This state of affairs will accentuate the loss of farmland to urban development. Between 2014 and 2022, urbanisation will gain more and more ground, nibbling away at periurban land whose main function was that of agriculture.

The following graph shows the evolution of the demand for planning permission between 2008 and 2019.

Figure 12: Change in the number of building permit applications from 2008 to 2019

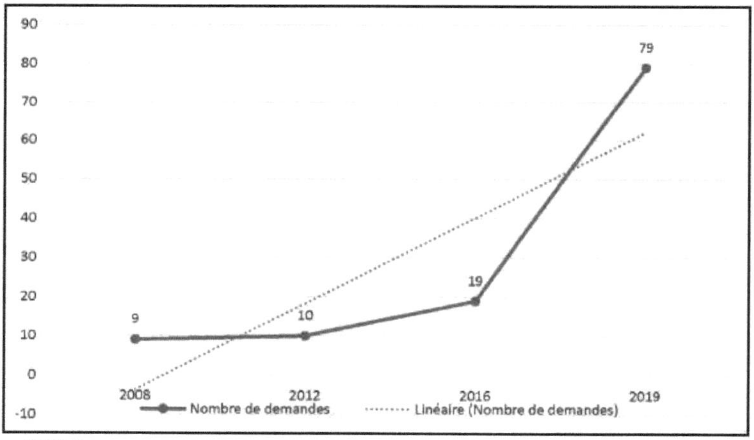

Source : DRUD

The graph shows that applications for planning permission have soared. In fact, the number of applications per year was no more than 10 in the 2010s. By 2019, however, it will have risen to 79.

I.2 Land speculation and growth in the property sector

In the past, the town of Bambey did not experience certain aspects of urbanisation, or even if it did, they had no major influence on its development. In fact, it was around 2010 that people from the rural communities of Ngoye, and more specifically from the village of Battal, made their way to the town of Bambey. Most of the requests for new buildings come from other communes, but especially from the village of Battal[32] .

Between 2011 and 2015, when the university began to operate at full capacity and its population grew, property became a preferred sector for most of the landowners around the town. The development of this sector is directly linked to the establishment of Alioune Diop University in the town of Bambey. The majority of building permit applications concern this part of the town next to the university (interview with the head of Bambey's town planning and housing department). However, most of the applicants come from the surrounding villages.

Today, this part of the city has seen the price of a plot rise, and a new urban centre is beginning to

[32] Gueye 2022, interview with the head of the Bambey town planning and housing department

emerge in this area, with the university at its heart. For a 300 m plot^2 , the lowest price can be estimated at between 2,500,000 F CFA and 3,000,0000 F CFA. Between 2008 and 2019, more than 90.60% of building permit applications reviewed by the town planning and housing department were for residential construction, according to the head of the Bambey town planning and housing department. Once built, these plots of land are usually rented out. It is mainly the villagers or people from surrounding villages who invest in the land, and are therefore the masters of the sector. Renting is gaining more and more ground as these landlords, who already have houses in their home villages, make their houses built in the town available to tenants and contribute to the rent boom. But in the town of Bambey, as everywhere else in the world, the price of rent depends on the neighbourhood. A district like Leona Nord, for example, has lower rental costs than the DVF district. In this part of the city, in the DVF-Diamaguene district near the university, the cost of rent is much higher, ranging from 30,000 to 40,000 CFA francs for single rooms and 35,000 to 45,000 CFA francs for rooms with a bathroom (interview with the head of the Bambey town planning and housing department).
This considerable increase in the price of rent can be explained by the distance-cost ratio. Indeed, given that the price of building materials such as cement and iron remains constant throughout the city, the only factor explaining the rise in the price of rent is distance from the university or comfort and quality. The alarming state of the city's transport sector, with inter-city mobility provided by carts, justifies the fact that rental values remain higher in the city centre districts. In the areas near the university and the CNRA, rental costs are becoming increasingly high compared with other areas.
The demographic dynamics of recent years are consuming space and endangering agricultural land (FAYE. M., 2013). The changing landscape in this area, marked by the development of modern buildings, shows that the city is spreading out mainly to the west, around the university, and to the south and south-east of the city, despite the high price of plots in this area. This speculation in land in the vicinity of the university is not without consequence, as it already generates a kind of spatial segregation. Indeed, given the excessively high cost of plots, poor families have no chance of obtaining a plot of land in this part of the city. There is already a form of wealthy neighbourhood compared to other parts of the city.
Those involved in land management, such as the municipality, have virtually left this land to the looters, without any regulation or anything. This state of affairs is harmful because there is still little control over land transactions, and the price of a square metre or a plot of land is skyrocketing. The law of the market is in full force because prices are consubstantial with supply and demand. However, the city's current strong demographic growth and future projections show that demand for housing land will continue to rise, given that the city's demographic growth is not currently subject to any constraints. In this part of the city, the median price of land to be developed is higher than in any other part of the city (FAYE M., 2013).
Despite the ever-increasing number of requests for land for housing, the town of Bambey finds itself in a very delicate situation with the extension of its municipal boundaries. As a result, the agricultural land of the surrounding villages is under threat, and the neighbouring communes are still hostile to the application of the town extension decree.
In the following graph, we will see the impacts of the city's spatial dëvelopment on land according to the enquëtës.
Figure 13: The effects of the city's spatial development on land

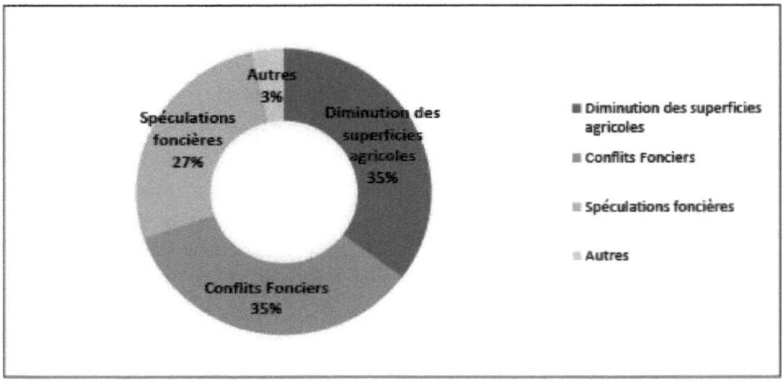

Source: Survey data, Gueye, October 2022

This diagram highlights the impact of urban sprawl on land. These include land speculation, a reduction in agricultural area, land conflicts and the inaccessibility of plots due to very high costs. The results of our survey confirm that urban sprawl in the city leads to land speculation (27%), a reduction in arable land on the outskirts of the city (35%) and land conflicts (35%).

I.2.1 Informal land transactions

Land is increasingly seen as a commodity, rather than a common good whose use must be carefully regulated in the interests of all citizens. This development has enormous consequences for the price of land, the way in which public land is used and the way in which its use is regulated by the State. All of these characteristics are affected by changes in urban requirements. So when the supply of land falls and demand rises, land prices will naturally rise, and fast. As a result, it is the wealthiest who will have access to the best land, rather than those who need it most. Market forces, demographic pressures and urban growth continue to drive up demand for land, but supply remains far below demand (UN-Habitat, 2011).

Current urban growth in the town of Bambey is having an impact on the land in the town and on the surrounding periurban land. The urbanisation of Bambey has given rise to a very dynamic land market. The town's strong demographic growth means that land is becoming increasingly scarce. As a result of the rush to find housing land, some of the farming communities in the surrounding communes have become involved in the land market. As a result, the price of a plot of land now varies between 1 and 5 million CFA francs. However, before 2005, a serviced plot of land measuring 300 m^2 could be obtained for less than 1 million (interview, 23 August 2022).

According to our survey results, land transactions in the town of Bambey have experienced a recent development dating back less than 5 years. Although the sale and purchase of land has existed in the town for more than 15 years, the sector had not experienced such a high level of growth and involved very few people. But at the moment, plot purchases are on the increase and the price has almost doubled, according to the head of the technical department at the town hall (interview, 23 August 2022).

Excessively high housing needs and a rapidly growing population mean that the price per square metre and the cost of a serviced plot are exorbitant. According to Ndiaye M., "*in the years 2003-2008, the price of a serviced plot was around 200,000 to 300,000 CFA francs. But these plots of land, which could be obtained at lower cost, are now being sold for between 7,000,000 and 8,000,000 CFA francs, because the migrant populations coming to settle in the town, as well as private individuals, are into property*" (interview, 23 August 2022). This rise in land prices can be explained on the one hand by the scarcity of land in the town. In fact, the densely populated urban centre and the partition between the RN3 and the railway, a space that is not occupied because it is considered to be a non aedificandi zone as it drains much of the rainwater, means that people are moving to the outskirts of the town. As a result, the demand for land in this area is outstripping supply, leading to an increase in the price per square metre.

The following table shows the evolution of land prices in the city.

Table 8: Situation of land transactions in the town of Bambey

Period	**Purchase price**			**Total**
		1 a million 5	**5 to 10 million**	
Minus 5 years	0	10	1	**11**
5 to 10 years	0	5	0	**5**
10 to 15 years	5	1	0	**6**
Over 15 years	4	0	0	**4**
Total	**9**	**16**	**1**	**26**

Source: Survey data, Gueye, October 2022

The table shows that 42% of land purchases took place less than 5 years ago. On the other hand, 61.54% of plots sold cost between 1 and 5 million euros. When the formal land market cannot provide enough land to meet the city's needs, land prices rise and a growing number of people will no longer be able to afford housing or land. And when the poor cannot afford land on the formal markets, they turn to the informal markets (UN-Habitat, 2011).

Given this situation on the land market, it is interesting to look at the socio-professional profile of those involved in land transactions in the city.

I.2.2 The price of plots of land, a factor in social segregation

The high price of plots of land in the town of Bambey and the surrounding area means that not all socio-professional groups can afford to buy a plot of land. In fact, the establishment of the university, which has given greater importance to the price of land in the surrounding neighbourhoods, suggests that this area is reserved exclusively for the wealthy social classes. The scarcity of land in the Bambey municipal area means that people have no chance of getting a plot at a lower cost.

According to our survey results, most land purchases in the city are made by civil servants and shopkeepers. More than 30.8% of the land purchased in the city is owned by civil servants, followed by shopkeepers, who often invest in property.

But what is most striking is the presence of farmers in the land market as buyers. In fact, people from the surrounding countryside are buying land to invest in property and becoming landlords. The proportion of land purchased by farmers exceeds 10%, reaching 11.5%. This justifies the statement by the head of the Bambey urban planning and housing department that rural dwellers are the masters of real estate (interview, October, 2022).

However, other social strata forced to look for a place to live are present on the land market. These social categories, grouped together in the 'Other' class or category, despite not having such great financial resources, play an important role in land purchases. This "Other" class includes carpenters, macon, mechanics and others working in little-known sectors. Like civil servants, they account for 30.8% of land purchases. This category also includes people whose secondary economic activity is real estate.

Table 9: Breakdown of socio-professional categories of land buyers

Socio-professional profile	**Percentage**
Farmer	11,5
Retailer	26,9
Civil servant	30,8
Other	30,8
Total	100

Source: Survey data, Gueye, October 2022

The urbanisation of the town of Bambey has had a number of impacts on land. They include the advance of the built-up area, land speculation marked by the high cost of plots and the development of real estate. However, uncontrolled urbanisation, which takes up a lot of space, is the key factor in this land situation, which in turn leads to land conflicts and disputes in the town.

Chapter II: Development prospects for the town of Bambey

Given the city's current demographic and economic dynamics, it has reached its territorial limits. However, the population continues to grow and the need for land for housing, infrastructure and facilities remains a major issue for the city authorities. This chapter looks at the development prospects available to the town of Bambey to meet its needs in a context where its master urban development plan has just come to an end. Before that, however, we have carried out an analysis of land requirements for housing, facilities and infrastructure.

II.1 Land requirements for the town of Bambey

Spatial planning also depends on changes in society and space (DIOP D. and TIMERA A. S., 2018). It is also, and above all, a kind of more or less long-term planning that affects the whole territory. Naturally, a multitude of decisions are needed to reorganise the territory[33] . The development of a given area requires a rational and hierarchical organisation of the spatial distribution of facilities and infrastructures. Thus, according to DIAKHATE M. M. (2011), development is carried out according to levels of demand and its scale proportional to the needs of the population in the area concerned.

If the reference urban development plan is to be believed, the areas to be urbanised in the town of Bambey for the period 2000 to 2010, with an average annual growth rate of 4.3% adopted as part of the preparation of the town's PDU, should reach a population of 30,908 in 2005 and 38,172 in 2010. This urban population situation implies a gradual improvement in the current economic situation of both the department and the city.

As a result of the Regional Development and Spatial Planning Programme, the town's population will increase from 25,047 in 2000 to 38,172 in 2010, representing an additional population of 13,125 people[34] . Estimates of land requirements for new extensions are based on this additional population.

According to the PDU for the town of Bambey, based on a gross density of 40 inhabitants/ha, a municipal area of 1,350 hectares will be needed to accommodate 53,956 inhabitants in 2021. Based on these estimates, an additional 675 hectares (including housing and facilities) will need to be developed to support the town's demographic and spatial dynamic. On the basis of 6 inhabitants per plot, which is below the national threshold (10 inhabitants per 150 m^2), the number of plots required to adequately accommodate the additional population is estimated at 4,500 plots. These current housing area requirements can be met by existing recent allotment projects with a capacity of 1,426 plots (DVF and Wakhaldiam).

Table 10: Estimated areas to be developed by 2021 according to the Bameby UDP

	2001	2006	2011	2016	2021
Population Total	26 993	32 357	38 337	44 993	53 956
Additional population		5 364	5 980	6 675	8 963
Space requirements (ha)	260ha (the city's official perimeter is well overflowing) 920 ha (draft municipal boundaries)	134	149	167	225

Source: PDU for the town of Bambey up to 2021

[33] Sueur, 2012. Cited by DIOP D. and TIMERA A. S. (2018)

[34] Reference Urban Development Plan for the town of Bambey

II.2 The town's infrastructure and equipment requirements

The town's current demographic weight 33,716 inhabitants according to ANSD projections for 2018, 38,239 inhabitants in 2022. These figures reveal a population that is constantly growing. As a result, the city's authorities are having to try and cope by finding mechanisms to meet the need for housing, schools, health facilities and other basic socio-cultural infrastructure. Estimates made on the basis of the 2001 UDP population projections show that 24,625 hectares of land should be found to meet these needs. This land area should be divided between the construction of school, health, shopping and transport facilities.

The table below shows the areas planned under the city's PDU for 2021.

Table 11: Facilities and surface areas planned under the PDU in 2021

EQUIPMENT	AREA IN HECTARES
School equipment	7.125 ha
Sanitary equipment	10 ha
Merchant equipment	0.2 ha
Sports and socio-cultural facilities	7 ha
Transport equipment	0.3 ha

Source: PDU de la ville de Bambey a l horizon 2021

The table shows that more than 40% of the area earmarked for facilities and infrastructure is earmarked for health facilities. This proves the city's real need to strengthen its medical facilities. The town of Bambey should have a hospital within the standards, given its importance in the department and its catchment area.

In terms of infrastructure, the town's road network is mainly made up of the RN3 Dakar-Diourbel, which provides access to the outside world, and regional roads such as the RN3 and RN32 (Fatick-Bambey-Meckhe), as well as the D306 departmental road leading to the commune of Gawane. This entire road network is linked by the ring road that surrounds the town of Bambey.

The road not only provides physical and functional links between the various centres and neighbourhoods, but also facilitates links between the city of Bambey and its hinterland on the one hand, and between Bambey and the rest of its territory on the other. In terms of their importance, the primary roads will have a total length of 40 metres. However, narrowing may be tolerated in places where it is virtually impossible to widen. In addition to the 5,500 linear metres (ml) of existing roads which are in good condition, 14,000 ml of new primary roads are planned for the peripheral network and the improvement of asphalting, as well as the tracing of 6,230 ml of primary roads (RN32 and RN31 and the D306).

Hierarchy 2 and 3 roads, which form a network of different parts of the urban fabric, provide internal links for inter-urban mobility. For these two categories of lanes, there are 11,537 ml of new secondary lanes and 10,367 ml of tertiary lanes, supplemented by 2,450 ml of rehabilitated tertiary lanes. The hierarchy 2 lanes will have a 20-metre right-of-way and will be supplemented by minor service roads with a right-of-way of between 10 and 15 metres. These are essentially sandy roads.

The upgrading of the various junctions and level crossings will go hand in hand with the upgrading of the road network[35] .

II.3 Constraints on the expansion of the town of Bambey

Development plans to enable Bambey to contain the additional population have come up against numerous constraints. The city's urban development master plan, which was due to be completed in 2021, has come to an end, or even exceeded it, and yet to date nothing has been done to extend the municipal boundaries of Bambey, apart from the two housing development projects that were due to extend the DVF and Leona Nord neighbourhoods in 2020.

As the city's basic urban planning document, the PDU, although coming to an end, deserves to be

[35] PDU for the town of Bambey up to 2021

updated, as the population estimates are no longer up-to-date, and as a result, the estimates of the need for housing space are also outdated. A comparison of the PDU's population estimates for 2021 and the ANSD's estimates clearly shows that the city is not yet ready to reach the 50,000 inhabitants in 2021 that the PDU forecast. In 2018, according to ANSD projections, the city's population stood at 33,716, rising to 38,239 in 2022. This figure should be exceeded if we go back to the PDU forecasts, which already estimated the city's population at 44,993 in 2016.

The opposition of the other local authorities to all attempts to incorporate their land into the Bambey municipal land base leaves the town uncertain in the long term. In fact, the authorities of the commune of Bambey relied on decree no. 2011-1991 of 16 December 2011 amending and extending the territorial boundaries of the commune of Bambey, but this decree is still being contested by the communal authorities of Ngoye, Thiakhar, Ndangalma and Ngogom.

Today, the town's development prospects are out of date, as legal and socio-cultural constraints leave the town authorities no hope of applying the decree. According to the head of Bambey's town planning and housing department, intercommunal cooperation between the four neighbouring communes is the only way to meet the town's current land needs.

Today, the situation in the town of Bambey remains alarming and worrying. Faced with the exhaustion of its communal boundaries and the many constraints on its urban expansion, the town of Bambey is becoming the scene of land disputes and conflicts. The increase in population is creating a need for housing and space for the construction of basic infrastructure and social facilities to support the town's growth. But the city is unable to provide housing for its population if nothing is done by the State or the city's local authorities.

In order to meet the expectations of its population, Bambey nevertheless sees itself as having a solution to this situation. That solution could be inter-municipality.

II.4 Inter-municipality, the solution for controlled urban growth

Inter-municipality allows communes to come together within a public body, either to provide certain services such as refuse collection, sanitation, urban transport, etc., or to draw up genuine economic development, development or town planning projects (INSEE definition). On the basis of this definition, the town of Bambey would have greater opportunities if it were able to join forces with the four neighbouring communes of Ngoye, Thiakhar, Ngogom and Ndangalma to form an inter-communal organisation with the aim of controlling its sustained urban growth and meeting the housing needs of its ever-growing population, for which the land reserves of these neighbouring communes seem to be the only ones capable of making up for such a shortfall. Most of the available land reserves around the town belong to the former rural communities. As for the town of Bambey, it is obliged to remain within its initial, already exhausted, communal boundaries. The problem of the administrative decoupage of the administrative districts therefore becomes a source of problems. Thus, NGOM K. F. (2019) argues that the decrees creating the collectivities merely list the villages attached without any other geolocation element. As a result, the territorial boundaries between collectivities become blurred and pose huge problems, most often between communes, or between former rural communes and urban communes.

The rural communities whose land base was once used as an extension zone for the urban communes. This is why (NGOM K. F., 2019) states: '*The former rural communities are the natural territory for the extension of the large border communities confined within tight territorial limits, most of which have used up their land base. The other type of direct conflict is that between former rural communities, as soon as one expresses a need or makes a decision on a plot of land close to the border"*. Nowadays, however, since the introduction of Act III of decentralisation, large urban communes such as Bambey have been faced with problems of expansion, as their very small surface area has already been exceeded and the land of the former rural communities, which was used as an expansion zone, has become a fiercely protected commercial asset. As a result, the rural communities are becoming hostile to any idea of ceding part of their area to another local authority. As a result, inter-communal conflicts abound and are becoming widespread throughout the country (NGOM K. F.,

2019)[36] . This leads BA B. and BA R. (2021) to say that the first fruits of Act III are not perceptible in the field of organising the territories or basic units that should support the implementation of development actions. The initial plan was to reframe territorial boundaries in order to provide a better framework for the dynamics taking place within them and to ensure territorial coherence. This measure has been slow to take shape, and in the meantime, the imprecision of the boundaries of newly communalised rural communities is fuelling recurring conflicts between neighbouring communities[37] .
The city's sustained demographic growth and its attractiveness have meant that it has continued to expand spatially since its birth. Today, however, it has to be said that the town is having problems containing its population, especially in the coming years. In fact, this explains the land tensions that have been evident in the town for some time now.
The expansion of the town has led to duality between the town and the surrounding communes. To resolve this situation, the city of Bambey and the four neighbouring communes need to join forces to form an inter-municipal organisation, in order to control the city's urban growth and avoid land insecurity around the surrounding farmland.
In Senegal, cooperation between local authorities predates law no. 96-06 of 22 March 1996 on the Local Authorities Code. In fact, it dates back to 1983 with the first attempt to bring together the municipalities of Dakar, Pikine and Rufisque (the Dakar urban municipality created by decree no. 83-1131 of 29 October 1983) around a waste management initiative. In legislative and regulatory terms, inter-municipality was created by the law referred to below, in which two articles were devoted to the possibility for local authorities "to undertake cooperative actions between themselves" (Article 14), as well as the possibility "to carry out programmes of common interest" (Article 15). It was from this date that local authorities acquired the legal capacity to engage in sectoral cooperation actions falling within their areas of competence in interregional entities (Article 71), mixed groupings (Articles 74 and 75), inter-municipal agreements (Article 179), urban communities (Articles 181 to 190) and Community Interest Groupings (Article 239 and 240).
As part of Act III of Decentralisation, Law no. 2013-10 of 28 December 2013 on the General Code of Local Authorities sets out the legal framework for inter-municipality in its relevant provisions, in Articles 16, 17 and 18, as a form of cooperation between local authorities.
Within the framework of respect for the principle of free administration, the State guarantees and organises the principle of solidarity between local authorities (articles 279 and 283, 285 and 291 of the General Code of Local Authorities), which may undertake programmes of common interest with the State. Thus, in terms of regulatory and legal provisions, inter-municipality becomes a form of organisation and joint governance based on strong leadership from local executives. Not only is it an opportunity to be seized, but it is also an approach based on solidarity and a sustainable way of managing structuring projects and local resources. It offers a typical model for large-scale territorial governance, with a view to ensuring that the problems encountered are adequately addressed. These issues include the exhaustion of the initial communal boundaries of urban communes affected by significant population growth and extraordinary land pressure. There are several forms of inter-municipality. But the most common form is that of association. The communes come together around a common interest and are bound by an agreement for a limited period[38] .
Intercommunality is based on six (6) solid principles[39] in order to better guarantee the independence of local authorities that take the initiative of grouping together in an organisation to manage a common problem. These principles include :

[36] NGOM K. F., 2019. *Acte 3: 4 plaies, 5ans apres!* consulted on 01 January 2023 on https://.seneplus.com/opinions/acte-3-4-plaies-5ans-apres
[37] BA B. and BA R., Juillet, 2021. *Acte III de la decentralisation senegalaise : l'epreuve de la coherence territoriale et du developpement territorial,* consulted on 02 January 2023 at http://ipar.sn/Acte-III-de-la-decentralisation-senegalaise-l-epreuve-de-la-coherence.html
[38] adl.sn/article/intercommunalite, jeudi, 06/08/2015, consulted on 01 january 2023
[39] GRDR, Guide de l'intercommunalite, October 2019. *L'intercommunalite, une reponse aux enjeux du développement local dans le bassin versant du fleuve du Senegal,* 98 p.

The principle of speciality is based on a dual speciality (formal and territorial). In terms of formal speciality, the EPCI has no general powers. It only has powers of attribution and only intervenes in the areas for which it is responsible. Territorial speciality refers to the fact that the organisation within which the various local authorities are grouped cannot act outside its territory. Thus, the scope of a group's powers or actions is limited to the territory of the local authorities it brings together.
The principle of exclusivity refers to the fact that a transfer of powers to the intermunicipal structure by one of the member municipalities results in the latter's relinquishment of responsibility for the said powers. In this case, the intercommunal structure becomes the only one with the power to act in the areas of competence that have been transferred to it. However, this principle does not preclude the transfer of certain areas of responsibility to another public entity (mixed syndicate, metropolitan pole, etc.). Furthermore, the principle of exclusivity does not prevent the division of a responsibility when it can be divided, but this cannot lead to the splitting of investment and operating operations within the same responsibility.
Thus, when a municipality relinquishes a competence to a EPCI, it cannot decide to entrust it to another EPCI without first withdrawing it from the previous EPCI.
Furthermore, the principle of exclusivity does not preclude the division of powers where it is appropriate, but this may not lead to the splitting of investment and operating operations within the same power.
Thus, when a commune relinquishes a responsibility to the intermunicipal structure, it cannot decide to entrust it to another EPCI without first withdrawing it from the previous EPCI.
The principle of reciprocity and solidarity is based on the fact that inter-municipality must serve the collective interest, that of the population of the entire inter-municipal territory and not that of a single commune or certain elected representatives.
The principle of equal responsibility, i.e. no commune or local authority may establish or exercise guardianship over another.
The principle of voluntariness on the part of the communes means that, when setting up an intercommunal structure, each commune must freely decide whether or not it wishes to join. It cannot therefore be forced or coerced into joining the intercommunal structure.
The principle of subsidiarity: in this case, the municipal level must be given priority for as long as the inter-municipal level cannot act effectively. For each competence transferred, there must be a strict and very clear delimitation of the fields of intervention of the intercommunal body. In other words, it is necessary to define what is in the Community interest and what remains the responsibility of the member municipalities.
There are many challenges associated with inter-municipality. It is a remedy for the fragmentation of local government and an instrument for the rational organisation of territories. Intercommunal cooperation brings together scattered resources and structures local initiatives. It encourages local economic development and revitalises regional planning policy. It is the establishment of a territorial network that makes it possible to respond to the challenges posed in terms of regional planning, whether these relate to the problems of urban development or the devitalisation of rural areas. It is a pragmatic response to management problems.
The competences exercised by the communaute de communes include: economic development; protection and enhancement of the environment, sustainable development (waste management, etc.), housing and living environment policy, sports and cultural facilities, primary and elementary education facilities, social action, sanitation, tourism, transport and town planning documents.Despite the progress made in terms of the legal and regulatory framework, it has to be said that inter-municipality is still marginal in view of the major constraints it has to face, including :
J The weakness and lack of clarity of the organisational and legal framework governing and organising inter-municipality initiatives;
The lack of political support for inter-municipality among elected representatives and local decision-making bodies;

Lack of knowledge and understanding of the challenges of inter-municipality ;
The strong technical and financial dependence of local authorities;
The lack of accurate statistical data on local authorities and of appropriate planning tools for existing forms of inter-municipality;
The lack of capacity on the part of elected representatives and other local players to build and implement inter-municipalities.

Broadly speaking, the town's increasing urbanisation is a real source of problems. Urban development is gaining ground, nibbling away at the agricultural land of the surrounding villages, because the town of Bambey has already exhausted its communal boundaries. As a result, the town's new development prospects are projected onto the land reserves of the surrounding communes. But today, its former rural communities are totally hostile to any attempt to transfer a portion of their land to the urban commune of Bambey. In 2020, this state of affairs led to tensions between the local authorities involved, following the initiation of housing development projects by the town of Bambey on farmland located on the outskirts of the town. However, it should be noted that the biggest problem is that of blurred territorial boundaries, which are difficult to understand on the ground because it is hard to see where the boundaries of one local authority or another end. So, hello tensions and blocked development projects. In addition to this, it should be noted that land disputes in the town also pit the residents against the municipality, with customary practices laying down the law in some of these neighbourhoods, commonly known as the '*Baay Deuk*'. Faced with such a situation, all attempts by the town of Bambey to improve the situation are in vain, and the solution to such a problem seems to be inter-municipality, despite the fact that the town's local authorities have not yet taken any steps in this direction. It now remains to be seen whether the local elected representatives of the town's neighbouring communes and those of the town of Bambey will manage to set up such a structure in the future to ensure coherent territorial organisation, controlled urban development and sustainable land management for the benefit of the local population.

The narrowness of the town's municipal boundary and its significant demographic growth since 1976 mean that today the town has reached the limits of its initial municipal boundary and is overflowing in all directions, nibbling away at the surrounding farmland on the outskirts of the town. Faced with such a situation, land is becoming a major issue for local authorities. But the stakes when it comes to land are different, depending on the nature and situation of these parts of the State. While the challenge for the commune of Bambey following the 2011 decree amending and extending the territorial boundaries of the commune of Bambey was to gain land reserves and provide housing for its ever-growing population, for the former rural communities of Ngoye, Ndangalma, Ngogom and Thiakhar the challenge is to protect the land of their populations, which supports economic activities, particularly agriculture. Over and above these issues, which vary from one commune to another, the former rural communities remain determined to thwart the decree, which would deprive them of part of their land, which is also a major source of funding for the local authorities, with the demarcation fees and taxes levied on the land.

GENERAL CONCLUSION

The town of Bambey began to take shape from the hamlets around 1986, and was granted full municipal status in 1958. Thus, since the 1960s, the town has continued to receive flows of people due to its geographical position, but also its attractiveness and, consequently, the development of facilities in the town.

As the only urban commune in an entire department of the same name, the town of Bambey has command powers over the rest of the department. This is justified by the town's political, administrative and economic functions, which in turn influence its dynamism. A dynamic town from the outset, it was in the 1970s, a period of drought, famine, misery and hunger that the town experienced its first major population influxes in its history. At that time, the countryside was repulsive and unliveable because of the difficult living conditions exacerbated by low agricultural yields, the basic economic activity of the rural world and a source of income for rural dwellers. As a result, the city is becoming the centre of refusal par excellence, because of the many opportunities it offers its population. These include education, employment and well-being. As a result, Bameby is becoming the preferred location for people from rural areas, under the famous concept of the rural exodus. This additional population, coupled with the town's sustained demographic growth over the last four population censuses, has meant that Bameby has become the place of choice for people from rural areas, under the famous concept of the rural exodus. For example, 16,268 inhabitants in 1988 (RGPH, 1988) to 20,366 inhabitants in 2002 (RGPH, 2002), i.e. an additional population of 4,098 inhabitants, then a population of 28,908 inhabitants at the time of the last census in 2013, i.e. an increase of 8,542 inhabitants or double the additional population of 2002, the town of Bambey then saw its spatial development become increasingly significant towards the years 2005. In fact, the years 2004 - 2005 were when the CUR was established in the town. The CUR, which became Bambey's Alioune Diop University in 2007, is a real factor in urbanisation. As a result, the town's spatial development is subject to significant demographic growth and the development of infrastructure and facilities such as the health centre and the town's central market. The latter is one of the structuring points for city-countryside relations between Bambey and the surrounding villages. This is because the town's central market is the largest in the whole department. The existence of a weekly livestock market that draws in all the traders from the markets at Dahra in Linguere and Mbafaye in the department of Mbour, among others, means that the town owes much of its dynamism to this market facility. At the end of the rainy season, it is the outlet for the harvests of farmers in the communes of the department. Of all the factors fuelling the town's economic dynamism and development, one stands out as the most important. It's the university. As the focal point of a human population, it is a major determinant of urban growth or urban sprawl. Before it was built, the outlying areas of the city were very accessible for housing needs, but since 2009, there has been a huge amount of speculation in land, which in turn has led to land insecurity and a rise in the price of plots. The high cost of land in the city, or even in the outlying areas, is due partly to the exhaustion of the city's municipal boundaries, and partly to the lack of control over the land market on the part of the authorities.

This situation gives rise to recurring land conflicts involving a wide range of players. There are two types of land conflict, depending on the players involved. Firstly, there are land conflicts between various local authorities (the town of Bambey and its neighbouring communes). This category of land dispute or problem arises from a very contiguous administrative division in which it is difficult to understand the territorial boundaries of the local authorities, but also from political initiatives on the part of the local authorities. This kind of problem was noted in the town in 2020, when the municipal authorities wanted to subdivide land on the outskirts of the town, and the former rural communities of Ngoye and Thiakhar, which were mainly concerned, expressed clear opposition to these projects. It should be noted, however, that the subdivision projects that the Bambey municipality wanted to initiate concerned land that had been included in the town's land base following decree no. 2011-1991 of 16 December 2011, which extended the town's territorial boundaries. However, the main argument

put forward by the municipal authorities of the other communes was that the decree had not been published in the official gazette and was therefore not enforceable against third parties.
Faced with these problematic situations, it seems necessary to focus on the essentials for the good of all local authorities in an associative or cooperative organisation within the framework of inter-municipality for coherent territorial organisation and harmonious urban development. Today, as the city of Bambey and the other communes face a growing number of problems, the stakes are becoming ever higher in terms of land management. While the city's local authorities have to take initiatives to meet the land requirements needed to support the city's growth, the local authorities in the former rural communities are in a strong position that they want to consolidate while securing their land. The other issue here is that land has become an important source of funding for local authorities in the context of property tax, and no local authority wants to see itself deprived of such an important funding resource.
To arrive at these results, we had recourse to documentation in our work in order to gain an insight into the dynamics of urban spaces and the problem of urban land. This study also required a methodology, and as part of the fieldwork we established a sample based on the number of households in each district of the city. We therefore selected a sample of 2% in proportion to the number of households in each district.
The various tools used in this study helped to achieve the objectives and confirm the initial hypotheses, while providing answers to the research questions.
Our first hypothesis has been confirmed, given that the urban dynamic of the town of Bambey is subject to demographic growth and the development of facilities in the town. This is the main cause of the town's spatial expansion, and therefore of the exhaustion of the town's communal perimeter.
According to the research results, after asking the question, it is safe to assume that the urban growth of the town of Bambey has given rise to multiple land issues. This has led to a lack of control over the land market, increased land speculation and led to land disputes because of the different interests that local authorities are defending. These research findings confirm our second hypothesis. These land-related issues give rise to land disputes between municipalities and between municipalities and the local population. This makes land management difficult in the area, as the interests of some differ from those of others.
At the end of the last hypothesis, which consisted of looking at the impacts of the city's spatial dynamics on periurban land, we can see that urban growth has many impacts on land. These include the loss of agricultural land, rising land prices, increased rental costs and rental values, as well as land conflicts. All of which goes to show the impact of the spatial dynamics of the town of Bambey on land in the surrounding communes. So it's easy to see that the upsurge in land disputes in the area is largely due to the urban growth of the town of Bambey, which has become a land-grabber.
In summary, we have outlined the various aspects of the spatial dynamics of the town of Bambey and the consequences of this spatial evolution. Aware of the potential shortcomings of this study, it deserves to be expanded and opened up to other urban areas in the future.

BIBLIOGRAPHY

General and specific works

- BAILLY A. and BEGUIN H., 2016. *Introduction a la geographie humaine,* Paris, Armand Colin, 9e edition, 240 p.
- BAILLY A., 1998. *Les concepts de la geographie humaine,* Paris, Armand Colin, 4e edition, 333 p.
- BANZO M., BOUJU S. et al. 2000. *Introduction a la geographie,* Hachette, 190 p.
- Beaujeu-Garnier J., 1997. *Geographie urbaine*, Paris, Armand Colin, 5e edition, 349 p.
- BRUNET R. et al, 2005. *Les mots de la geographie, dictionnaire critique,* Reclus-La Documentation Fran^aise, 3e edition, 518 p.
- BURGEL G., 2017. *Questions urbaines,* Paris, Editions de l'Aube, 144 p.
- CHALEARD J. L. and DUBRESSON A., 1999. *Villes et campagnes dans les pays du Sud*, Editions Karthala, 258 p.
- CIATTONI A. and VEYRET Y., 2013. *Les fondamentaux de la geographie*, Paris, Armand Colin, 318 p.
- DERRUAU M., 2002. *Geographie humaine,* Paris, Armand Colin, 8e edition, 448 p.
- DIAKHATE M. M., 2011. *L'amenagement du territoire au Senegal: Principes, pratiques et devoirs pour le XXIe siecle,* Paris, L'Harmattan, 251 p.
- DIOP D. and TIMERA A. S., 2018. *Diamniadio: Naissance d'une nouvelle ville: enjeux et defis d'une gouvernance durable*, Dakar, L'Harmattan-Senegal, 228 p.
- DIOP D., 2012. *Urbanisation et gestion du foncier urbain a Dakar, Defis et perspectives,* Dakar, L'Harmattan-Senegal, 319 p.
- FALL A. S., GUEYE C. et al, 2005. *Urbain-rural, l'hybridation en marche*, Dakar, Enda Tiers-monde, 480 p.
- LE BRIS E. et al, 1982, *Enjeux fonciers en Afrique Noire*, Paris, Karthala, 431 p.
- LEFEBVRE H., 1970. *Du rural a I'urbain*, Paris, Antropos, 287 p.
- MANGA C. T., (under the dir. of), 2012. *Le Senegal, quelles evolutions territoriales,* Paris, L'Harmattan, 308 p.
- MBAYE I., (ed.), 2018. *La recomposition des espaces urbain et periurbain face aux changements climatiques en Afrique de l'Ouest,* Dakar, L'Harmattan-Senegal, 300 p.
- MERLIN P. and CHOAY F., 1998. *Dictionnaire de l'urbanisme et de l'amenagement*, Presse universitaire de France, 2e edition, 863 p.
- OSMONT A., 1995. *La banque mondiale et les villes, du developpementa l'ajustement,* Paris Karthala, 309 p.
- PAULET J. P., 2006. *Geographie urbaine,* Paris, Armand Colin, 342 p.
- PAULET J. P., 2009. *Manuel de geographie urbaine,* Paris, Armand Colin, 300 p.
- RAYNAUD M. M., DIOP D. et al., 2013. *Repenser les moyens d'une securisation fonciere urbaine. Le cas de l'Afrique francophone,* Canada, Service d'Impression de I'Universite de Montreal (SIUM), 223 p.
- TEULON F., (ed.), 2010. *Dictionnaire d'histoire, economie, finance, geographie*, Presses Universitaires de France - PUF, 6e edition, 752 p.

These, dissertations and scientific articles

- BA A., 2015. *Dynamique d'urbanisation de la ville de Thies,* master's thesis in geography, Universite Cheikh Anta Diop de Dakar, 68 p.
- CASSE L. O., 2010. *La production fonciere et immobiliere dans les quartiers péripheriques de Mbour,* Master's thesis in geography, Cheikh Anta Diop University, Dakar, 147 p.
- CASSE L. O., 2017. *Les filieres foncieres et immobilieres : de la production a l'usage sur le littoral Somone-Mbour de la petite cote senegalais,* PhD thesis in geography, Universite Cheikh Anta Diop de Dakar, 468 p.

- CLAVEL M., 2018. *L'amelioration de la gouvernance fonciere dans la zone d'amenagement hydro-agricole de la vallee du fleuve Senegal : Une contribution par l'analyse de genre,* Master's thesis, Universite Toulouse Jean Jaures, 107 p.
- DIENG A., 2018. *L'impact sur les terres agricoles dans la commune de Sindia,* master's thesis in geography, Universite Cheikh Anta Diop de Dakar, 70 p.
- DIOP A., 2005. *Ville et amenagement du territoire au Senegal,* doctoral thesis in geography, Cheikh Anta Diop University, 400 p.
- FAYE A., 2013. *Migration internationale et urbanisation : l'apport des emigres dans le développement de la ville de Bambey,* master's thesis in geography, Universite Cheikh Anta Diop de Dakar, 102 p.
- FAYE M., 2014. *L'implantation de l'universite Alioune de Bambey comme facteur d'urbanisation*, master's thesis in geography, Universite Cheikh Anta Diop de Dakar, 85 p.
- FAYE S., 2018. *Dynamique spatiale et les enjeux de l'amenagement du territoire : cas de la ville de Bambey,* master's thesis in geography, Universite Cheikh Anta Diop de Dakar, 110 p.
- MBOW D., 2014. *Croissance urbaine et enjeux fonciers a Dakar : cas du titre foncier 1975/R Bambilor,* master's thesis in geography, Universite Cheikh Anta Diop, 82 p.
- NDIAYE E. M., 2013. *Le secteur des transports urbains a Dakar : Problematique de la gestion des gares routieres et de la mobilite urbaine,* PhD thesis in geography, Universite Cheikh Anta Diop de Dakar, 517 p.
- NIANG F., 2020. *Mutations urbaines, dynamiques sociodemographiques et modernisation a Dakar : vers une transition des modes de déplacements,* Master's thesis in urban planning and development, Ecole d'Urbanisme de Paris, 131 p.
- ORFEUIL J-P., 2010. "Mobility and travel problems in the world's cities", *Hors-Series*, n°447 : pp. 68
- OUATTARA S., 2007. *Les enjeux fonciers : les stratégies de preservation et d'appropriation dans le département de Diebougou (Province de la Bougouriba) Burkina Faso ?* These de doctorant en sociologie, Universite Cheikh Anta Diop de Dakar, 324 p.
- RAYMOND J. and CHALINE C., 1983. "La dynamique urbaine", *Annales de Geographies,* n°514 : pp. 712 - 714.
- TINE M. L., 2016. *Eaux et assainissements dans la commune de Bambey,* master's thesis in geography, Universite Cheikh Anta Diop de Dakar, 127 p.
- WADE C. S., 1995, *Saint-Louis: la crise de sa croissance urbaine recente,* doctoral thesis in geography, Cheikh Anta Diop University, Dakar, 274 p.

Official documents

- ANSD, 2008. *Situation Economique et Sociale de la region de Diourbel 2008,* 147 p.
- ANSD, 2013. *RGPHAE,* 67 p.
- ANSD, 2013. *Rapport definitif RGPHAE* 2013, 418 p.
- ANSD, 2014. *Situation economique et sociale regionale 2014,* 209 p.
- ANSD, 2015. *Situation economique et sociale regionale 2012,* 205 p.
- ANSD, 2016. *Situation economique et sociale du Senegal 2016,* 2010 p.
- ANSD, August 2006. *Situation economique et sociale de la région de Diourbel 2005,* 72 p.
- ANSD, August 2011. *Situation economique et sociale regionale de Diourbel 2010,* 193 p.
- ANSD, April 2015, *Situation economique et sociale regionale 2013,*
- ANSD, March 2019. *Situation economique et sociale regionale de Diourbel 2016,* 210 p.
- ANSD, September 2018. *Situation Economique et Sociale de la region de Diourbel 2015,* 201 p.
- ANSD, September 2021. *Situation economique et sociale regionale de Diourbel 2019,* 152 p.
- World Bank, 2015. *Villes emergentes pour un Senegal emergent,* 126 p.
- DPS, September 1992. *RGPH,* Diourbel regional report, 69 p.
- IPAR, December 2010 *Workshop on land management in Senegal: issues, state of play and debates,* 42 p.

- IPAR, December 2019. *Information booklet on : Les procedures d'acces et les droits d'usage sur les terres des domaines national et de l'Etat,* 34 p.
- MINISTERE DE L'INTERIEUR, 2022. *Mapping community land disputes in Senegal,* 107 p.
- MINISTRY OF HEALTH AND SOCIAL ACTION, 2019. *Tome 5 de la carte sanitaire : Etat des lieux de l'offre de soins de sante et besoins globaux,* 398 p.
- OECD/CSAO, 2020. *Dynamiques de l'urbanisation africaine 2020 : Africapolis, a new urban geography, West Africa Papers,* OECD Publishing, Paris, https://doi.org/10.1787/481c7f49-fr., 208 p.
- UN-HABITAT, 2011. *Housing the poor in African cities,* 51 p.
- Reference Urban Development Plan for the town of Bambey
- Urban Master Plan for the City of Bambey Horizon 2021, 41 p.
- Plan National d'Amenagement et de Developpement Territorial horizon 2035, June 2020
- Service Regional de la Statistique et de la Demographie de Diourbel, 2016. *Situation economique et sociale regionale*, ANSD, 210 p.

APPENDICES

Laws and decrees

- Decree no. 64-573 of 30 July 1964 setting the conditions for application of law no. 64-46 of 11 June 1964 relating to the Domaine National (French national estate)
- Decree no. 72-1288 of 27 October 1972 on the conditions for the allocation and disallocation of land in the national domain included in the amended rural communities
- Decree no. 2009-1450 of 30 December 2009 on the regulatory part of the town planning code
- Law n°64-46 of 17 June 1964 relating to the National Estate
- Law n°76-66 of 2 July 1976 on the State Property Code
- Law no. 96-06 of 22 March 1996 on the Local Authorities Code
- Act no. 2008-43 of 20 August 2008 on the town planning code
- Law n°2011-07 of 30 March 2011 on the land ownership regime
- Law n°2013-10 of 28 December 2013 on the General Code of Local Authorities

Survey questionnaire for heads of household

I. HOUSEHOLD IDENTIFICATION:

I. IDENTIFICATION DU MENAGE:

Localisation:

latitude (x.y °)

longitude (x.y °)

altitude (m)

précision (m)

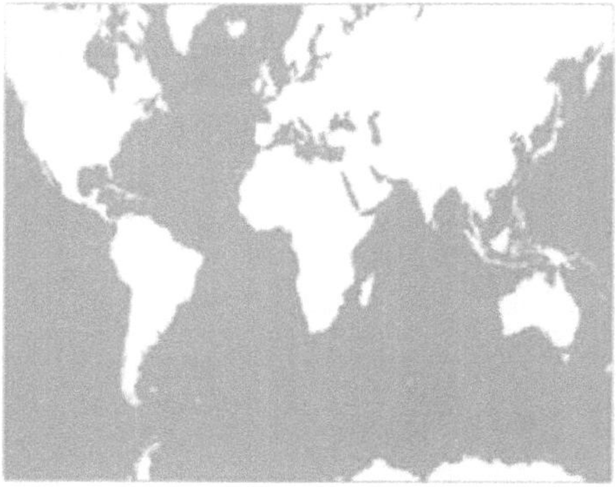

Nom:

Sexe:

- ◯ Masculin
- ◯ Féminin

Âge:

- ◯ 30 - 40 ans
- ◯ 40 - 50 ans
- ◯ 50 - 60 ans
- ◯ Plus de 60 ans

Profession:

- ◯ Agriculteur
- ◯ Commerçant
- ◯ Fonctionnaire
- ◯ Autres

Si "Autres", à préciser:

Avez-vous une activité économique secondaire?

- ◯ OUI
- ◯ NON

Si "OUI", à préciser:

____

Quartiers?

- ◯ DVF
- ◯ Wakhaldiam
- ◯ Léona Nord
- ◯ Léona Sud

Depuis combien d'années vous habitez dans ce quartier?

- ◯ Moins 5 ans
- ◯ 5 - 10 ans
- ◯ 10 - 15 ans
- ◯ Plus de 15 ans

Type de logement?

- ◯ Case
- ◯ Baraque
- ◯ Maison à basse
- ◯ Maison étage
- ◯ Immeuble
- ◯ Autres

Si "Autres", à préciser:

____

Mode d'acquisition du logement?

- ◯ Délibération de la mairie
- ◯ Achat
- ◯ Location
- ◯ Location/Vente
- ◯ Autres

Si "Autres", à préciser:

____

Statut d'occupation?

- ◯ Propriétaire
- ◯ Locataire
- ◯ Autres

Si "Autres", à préciser:

II. INFRASTRUCTURES ET EQUIPEMENTS:

Disposez-vous d'une structure de santé dans le quartier?

- ◯ OUI
- ◯ NON

Si "OUI", de quel type?

Centre de santé

Poste de santé

Case de santé

Maternité

Autres

Disposez-vous d'un établissement scolaire dans le quartier?

- ◯ OUI
- ◯ NON

Si "OUI", de quel type?

Crèche

Garderie

Elémentaire

Moyen

Secondaire

III. DYNAMIQUE URBAINE ET ENJEUX FONCIERS:

Pensez-vous que la ville de Bambey se développement spatialement?

- ◯ OUI
- ◯ NON

Si "OUI", quelles sont les raisons?

Croissance démographique rapide

Implantation de l'UADB

Position carrefour

Autoroute Ila Touba

RN3

Quels sont les effets de l'étalement urbain de la ville de Bambey sur les terres agricoles?

- [ ] Diminution des supercifies agricoles
- [ ] Baisse des rendements agricoles
- [ ] Spéculations foncières
- [ ] Conflits fonciers
- [ ] Autres

Si "Autres", à préciser:

Avez-vous acheté des terres ces dernières années?

○ OUI

○ NON

Si "OUI", il y a de cela combien d'années?

Moins 5 ans

5 - 10 ans

10 - 15 ans

Plus de 15 ans

Si "OUI", combien s'élève le prix?

Moins 1 million

1 - 5 millions

5 - 10 millions

Plus de 10 millions

Avez-vous vendu des terres ces dernières années?

○ OUI

○ NON

Si "OUI", il y a de cela combien d'années?

Moins 5 ans

5 -10 ans

10 - 15 ans

Plus de 15 ans

Si "OUI", à combien s'élève le prix?

Moins 1 million

1 - 5 millions

5 - 10 millions

Plus de 10 millions

Êtes-vous une fois confrontés à des litiges fonciers?

- ◯ OUI
- ◯ NON

Si "OUI", il y a combien d'années?

Moins 5 ans

5 - 10 ans

10 - 15 ans

Plus de 15 ans

Si "OUI", contre qui?

La mairie

Un privé

Un autochtone

Autres

Si "Autres", à préciser:

Comment vous voyez les impacts de l'UADB sur la ville de Bambey?

- ◯ Très négatifs
- ◯ Négatifs
- ◯ Très positifs
- ◯ Positifs
- ◯ Sans avis

Quelles sont les conséquences de l'implantation de l'UADB sur le foncier?

- [ ] Hausse des prix de la parcelle
- [ ] Spéculations foncières
- [ ] Pression sur le foncier agricole
- [ ] Perte d'espaces de cultures
- [ ] Conflits fonciers
- [ ] Autres

Si "Autres", à préciser:

Quelles sont les difficultés que vous pose l'implantation de l'UADB dans la ville de Bambey?

Quelles sont vos recommandations pour faire face à ces problèmes et que la ville de Bambey puisse profiter d'avantage des opportunités qu'offre l'UADB?

IV. MOBILITE ET TRANSPORT:

Quels sont les moyens de transport que vous utilisez fréquemment pour vos déplacements dans la ville?

- [ ] Charrette
- [ ] Moto
- [ ] Voiture
- [ ] Autres

Si "Autres", à préciser:

Comment vous voyez l'offre de transport dans la ville?

- ◯ Très mauvais
- ◯ Mauvais
- ◯ Acceptable
- ◯ Bien
- ◯ Très bien
- ◯ Sans avis

Quelles sont les difficultés que vous rencontrez en termes de mobilité?

Quelles solutions préconisez-vous pour améliorer ce secteur dans la ville?

V. ASSAINISSEMENT ET GESTION DES DECHETS:

Êtes-vous confrontés à des problèmes d'inondations?

- OUI
- NON

Existe-t-il un réseau d'assainissement dans le quartier?

- OUI
- NON

Êtes-vous confrontés à un problème de gestion des ordures ménagères?

- OUI
- NON

Existe-t-il un système de ramassage des ordures ménagères mise en place par la municipalité?

- OUI
- NON

Si "OUI", à préciser:

Si "NON", comment vous gérer ces ordures?

Quelles sont vos suggestions pour améliorer la gestion des ordures ménagères dans la ville?

Phots:

Cliquez ici pour téléverser un fichier. (< 5MB)

Interview guides

Interview guide for the head of the technical department of Bambey town hall:

1. How many years have you been in this position?
2. In what year was the town of Bambey founded?
3. How many neighbourhoods were there in the city?
4. How many neighbourhoods does the city currently have?
5. In what year were these new districts created and why?
6. How big is the population of Bambey?
7. How is this population distributed across the city?
8. What is the most densely populated neighbourhood and why is it so densely populated?
9. Which neighbourhood has the lowest population density and why?
10. How big was the town of Bambey before the year 2000?
11. How large is the town today?
12. What are the city's business sectors?
13. What is your main business sector?
14. What are Bambey's major facilities?
15. How do these facilities contribute to the city's growth?
16. Why has the city's surface area increased?
17. Where is the city's spatial expansion heading?
18. How would you characterise the city's current urban expansion?
19. What are the consequences of the urban sprawl of the city on the land of the surrounding communes?
20. Are there any housing development projects already completed or in progress?
21. How many hectares are involved, and how many plots are planned?
22. Is this land to be developed within the municipal boundaries of Bambey?
23. Which commune is most affected by urban sprawl in the town of Bambey?
24. On what basis was the project to extend the city's territorial limits based?
25. How have the surrounding municipalities reacted to this plan to extend the city's boundaries?
26. Are there any land disputes caused by this town extension project?
27. Who are the different players in these land disputes?
28. How do you manage these land disputes?
29. What role does the UADB play in land speculation?
30. What explains the rise in plot prices in the city?
31. What are the problems in terms of mobility and transport in the city?
32. What are your recommendations for developing transport in the city?
33. What are the sanitation problems in the town of Bambey?
34. What waste management policies are in place in the city?
35. What are your recommendations for better waste management in the city?

Interview guide for the Head of the Bambey Urban Planning and Housing Department

1. The spatial organisation of the town of Bambey?
2. Land situation in the town of Bambey ?
3. The current surface area of the town of Bambey?
4. What are the different phases in the city's urbanisation process?
5. What role does the university play in the city's urban growth?
6. What are the consequences of the UABD's establishment in the town of Bambey on the land market?
7. What impact is urban sprawl having on agricultural land in the surrounding villages?
8. What are the factors driving urban growth in the city?
9. The city's sanitation problems?
10. Waste management?

11. Given the town's rapid demographic growth, what are the space requirements for the construction of facilities and infrastructure, as well as residential space?
12. What are the constraints on these futuristic projects?
13. The resurgence of land disputes around the city: what are the causes?
14. Who are the various players involved in land management in the municipality?
15. Are there any documents (tools) for controlling the city's urban growth?
16. What role does the town planning department play in the town's urban development and land management?

Interview guide for members of civil society :

1. What are the factors behind the city's spatial development?
2. How do you see the effects of UADB's presence in the city?
3. What are the land issues in the city?
4. What is at the root of these land disputes in the town?
5. What do you think about families living in informal settlements?
6. What is your role as a movement to defend the land of Bambey?
7. What strategies have been adopted to deal with these land issues?
8. What is the role and place of the municipality in this land insecurity?
9. What do you propose as a solution to reduce land disputes in the city?

Map 10: Location of households surveyed

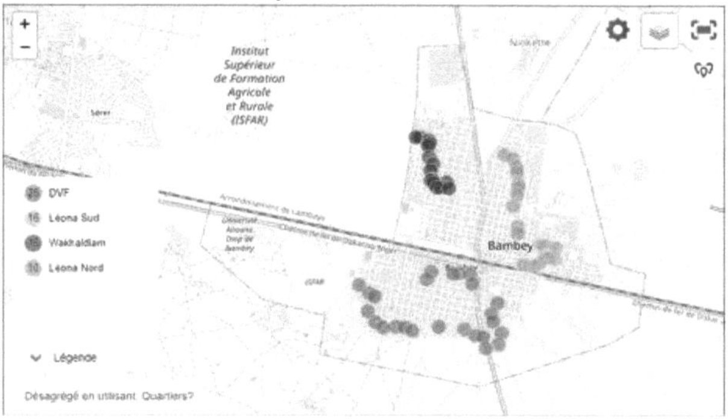

Source: OSM, Kobotoolbox, Guèye 2022 survey data **Figure 14:** Proportion of respondents by age group

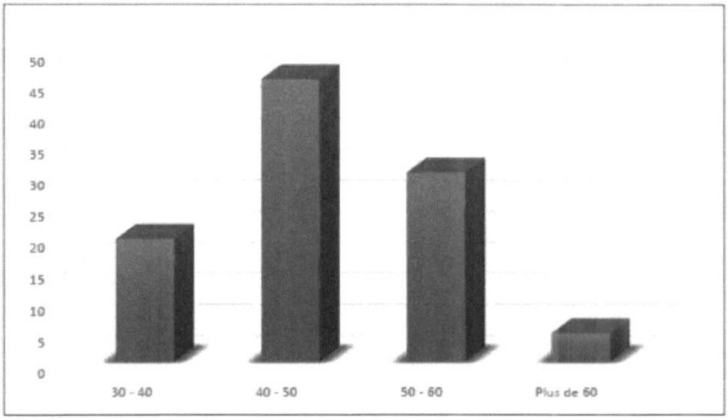

Source: Survey data, Guèye, October 2020
Figure 15: Socio-professional profile of respondents

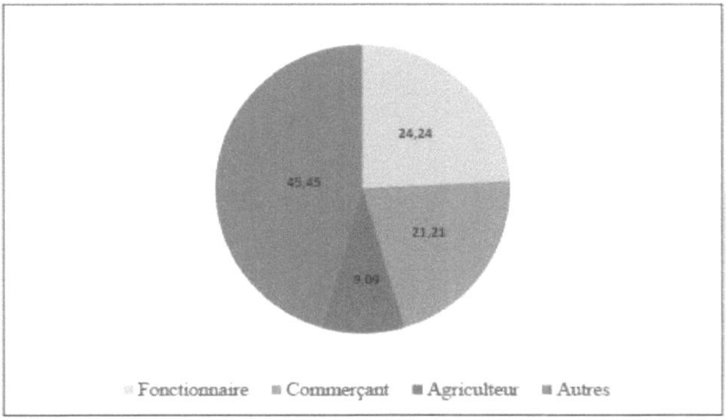

Source: Survey data, Guèye, October 2020 **Figure 16:** How respondents acquired their homes

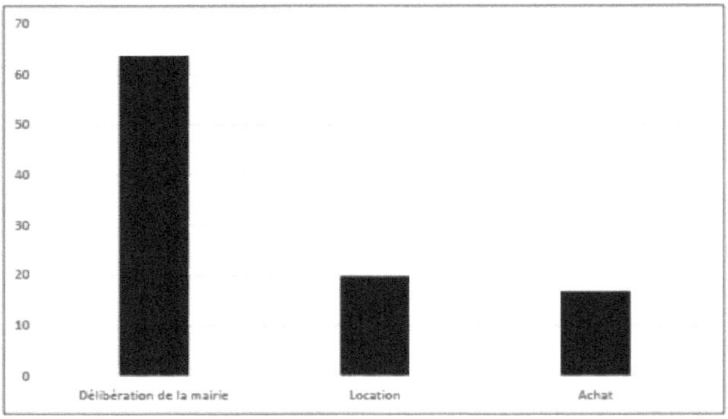

Source: Survey data, Guèye, October 2020
Figure 17: Tenure status of respondents

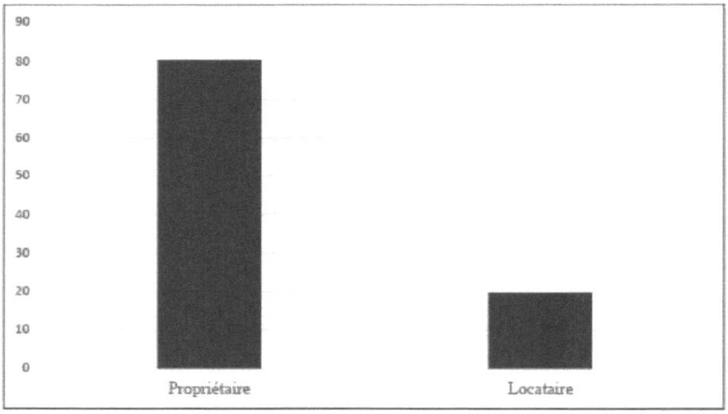

Source: Survey data, Guèye, October 2020 **Figure 18:** Type of house of respondents

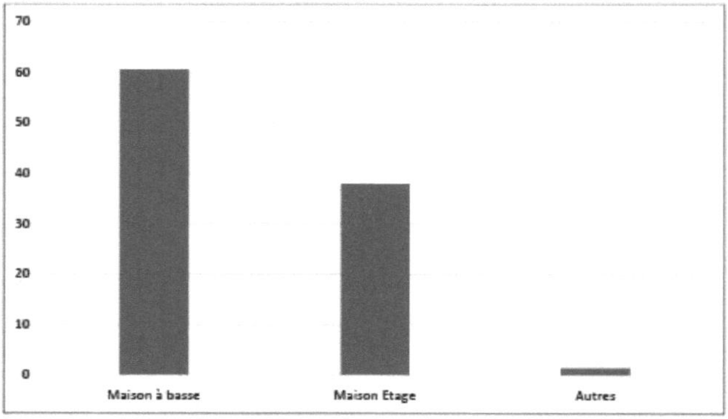

Source: Survey data, Guèye, October 2020

MIX
Papier aus verantwortungsvollen Quellen
Paper from responsible sources
FSC® C105338

Printed by Books on Demand GmbH, Norderstedt / Germany